Comprehensive Series in Photochemical and Photobiological Sciences

Volume 1

This book series offers a comprehensive overview of specific areas of photoscience and presents a critical evaluation of the trajectory that the field of photoscience is undertaking, covering selected aspects of the interaction between light and living matter. Committed to bridging the knowledge gap between established and emerging areas, this book series encompasses topics ranging from the photophysical events, followed by the absorption of photons by endogenous or exogenous constituents of biological systems to the large variety of effects caused by UV, visible, and near-IR wavelengths at a supramolecular, cellular and tissue level.

Particular attention is given to strategies for protecting ecosystems and plants, animals, and human beings against photodamage, as well as to using light as an information carrier or a beneficial tool in the field of medicine and the environment. An in-depth description of novel emerging topics, such as the use of nanotechnologies in photobiology or the advent of sophisticated time- and space-resolved spectroscopic and microscopic techniques, is also provided. By incorporating these emerging frontiers, the series provides a high-quality critical review of the state-of-the-art and presents future developments for each subject.

This book series supports and advances the UN's sustainable development goals, in particular, SDG7 (Affordable and Clean Energy), SDG3 (Good Health and Well Being) and SDG6 (Clean Water and Sanitation).

Initiated by the European Society for Photobiology, this book series expands the scope of photobiology and attracts the attention of the scientific and medical community to the high potential and value of this branch of science at both an academic, industrial, and social level.

María Gabriela Lagorio
Editor

Fluorescence of Living Plants

Basics, Applications and Trends

With contributions by

G. Agati · P. I. Calzadilla · S. Choza-Farias · V. Chukhutsina
G. B. Cordon · R. L. Gómez · G. N. Johnson · O. Knopf
A. Krieger Liszkay · D. Konche · M. G. Lagorio
I. López Valiño · O. Muller · M. Petri · F. Pinto · U. Rascher
J. M. Romero · B. Siegmann

Springer

Editor
María Gabriela Lagorio
Facultad de Ciencias Exactas y Naturales
University of Buenos Aires
Ciudad Autónoma de Buenos Aires, Argentina

ISSN 2041-9716 ISSN 2041-9724 (electronic)
Comprehensive Series in Photochemical and Photobiological Sciences
ISBN 978-3-032-04854-7 ISBN 978-3-032-04855-4 (eBook)
https://doi.org/10.1007/978-3-032-04855-4

This Springer imprint is published by the registered company Springer Nature Switzerland AG
The registered company address is: Gewerbestrasse 11, 6330 Cham, Switzerland

To my daughter, Daniela,
for the joy, light, and future she brings

To my husband, Daniel,
for his partnership on every step of the way

To my parents, María Luisa *and* Bruno,
whose memory continues to guide and inspire me

*To those who taught me to wonder,
and to those who never let me stop*

Preface

Plants do not speak … but through the light they emit as fluorescence, they express how they feel. They share information about their physiological state and the performance of their photosynthetic machinery.

Chlorophyll fluorescence in photosynthetic organisms is a *silent language of powerful expression*.Within that luminous signal lies a message: about their health, carbon assimilation, and their interaction with the environment.

The spirit of this book is to delve deeply into the scientific foundations behind that idea. With an integrative and didactic approach, it leads the reader from fundamental principles to the most advanced concepts currently available in the field.

This book is intended for researchers, professionals, and university-level students—both undergraduate and postgraduate—interested in chlorophyll fluorescence and its wide-ranging applications: from photosynthesis research and agricultural development to postharvest fruit and vegetable monitoring, satellite-based remote sensing, and carbon capture.

This volume is structured into eight chapters, each offering a journey through different aspects of chlorophyll fluorescence study and analysis. The general goal is to provide a clear, accessible path that guides the reader through some of the many directions one can explore within this fascinating field.

Chapter 1 (Chlorophyll Fluorescence: History) serves as an entryway into the subject, describing the historical developments that led to the discovery of fluorescence in general, and chlorophyll fluorescence in particular. This chapter offers interesting historical accounts, some of which trace back to the era of Spanish exploration and conquest in the Americas.

Chapter 2 (Chlorophyll Fluorescence and Photosynthesis) presents the fundamental principles linking fluorescence emitted by chloroplasts to the primary processes of photosynthesis. It provides a conceptual foundation necessary for understanding the more detailed discussions in the chapters that follow.

Chapter 3 (Non-photochemical Quenching in Plants: A Chlorophyll-a Fluorescence Perspective) explores in depth the mechanisms of heat dissipation known as non-photochemical quenching (NPQ). This regulatory process fine-tunes photosynthetic efficiency in response to light intensity, protecting the photosystems from excess excitation.

Chapter 4 (Time-Resolved Fluorescence Spectroscopy in Plants) presents the principles, methodologies, and key challenges involved in time-resolved fluorescence measurements in intact leaves. It highlights recent technological and conceptual advancements while also identifying current knowledge gaps in the field.

Chapter 5 (Proximal Remote Sensing: Simultaneous Active and Passive Measurements of Fluorescence at the Interface from Single Leaves to the Canopy) focuses primarily on remote and noninvasive measurement techniques and how these can provide valuable insights into crop physiology. It introduces key concepts related to fluorescence-based remote sensing.

Chapter 6 (From Space to Cells: Physical Models for Light Re-absorption and Scattering) explores physical models used to correct the spectral distribution and quantum yield of fluorescence, accounting for light re-absorption and scattering processes in both individual leaves and plant canopies.

Chapter 7 (Fluorescence Applications on Plants) outlines recent and emerging applications of chlorophyll fluorescence in plant science, spanning both research and practical uses.

Finally, Chapter 8 (Perspectives on Chlorophyll Fluorescence) summarizes the latest developments in the analysis of chlorophyll fluorescence, including its integration with data science and artificial intelligence for automated and robotic monitoring systems. It offers a forward-looking view of promising directions and future challenges in the field.

The completion of this book and the diverse insights captured in each chapter are the result of a collaborative effort involving many generous and talented individuals. I would like to take a space to acknowledge and thank those whose support made this work possible.

First and foremost, I would like to thank Dr. Massimo Trotta for trusting me with the editorial responsibility of this book.

Secondly, I wish to express my gratitude to Sofía Costa, Neelofar Yasmeen, and the entire team supporting Springer's publishing work for their assistance throughout the process.

Special thanks go to the authors, who joined me in the challenge and adventure of putting this volume together. I truly appreciate their willingness to contribute. This book simply would not have been possible without their valuable input.

I am also grateful to the European Society for Photobiology, the International Union of Photobiology, and the German Research Foundation (DFG) for their financial support, which enabled me to present my work at recent conferences and facilitated direct contact with leading professionals in the field.

I also extend my ongoing gratitude to the University of Buenos Aires, which shaped my academic path and, together with CONICET, continues to support my work as a professor and researcher.

I hope that this book serves as an inspiring and useful guide for those interested in the topic of chlorophyll fluorescence. I wish it sparks curiosity, ignites new ideas, and encourages further exploration in this exciting field.

Ciudad Autónoma de Buenos Aires, Argentina María Gabriela Lagorio

Contents

Chlorophyll Fluorescence: History

María Gabriela Lagorio

Abstract The history of fluorescence is closely connected to natural products. The first documented observation occurred in 1565 when a natural extract, used as a medicinal infusion for treating urinary tract infections, was noted for its distinctive "blue colour", later revealed to be fluorescence emission. This discovery is also connected to the conquest of the Americas and the traditions passed from Mexican natives to the Spaniards. In this chapter, we begin by exploring the main significant events in the history of fluorescence. We then focus on the history and evolution of chlorophyll fluorescence, whose analysis and interpretation are central to the book's theme. It is an engaging journey that illustrates how early researchers' insightful observations of luminous phenomena with the naked eye enabled the way for developing innovative methods to analyze chlorophyll fluorescence and its relationship to the photosynthetic process. These methods evolved from studying individual leaves to entire canopies and even remote observations via satellite monitoring of vegetation.

Keywords Brewster, Chlorophyll fluorescence, de Sahagún, Fluorescence history, Kautsky, *Lignum nephriticum*, Monardes, Stokes

Luminescence is "the spontaneous emission of radiation from an electronically excited species (or a vibrationally excited species) that is not in thermal equilibrium with its surroundings" [1]. In particular, the phenomenon of fluorescence is a type of

The original version of the chapter has been revised. A correction to this chapter can be found at https://doi.org/10.1007/17361_2025_13

M. G. Lagorio (✉)
Facultad de Ciencias Exactas y Naturales, Dpto. de Química Inorgánica, Analítica y Química Física, Universidad de Buenos Aires, Buenos Aires, Argentina

CONICET, Instituto de Química Física de los Materiales, Medio Ambiente y Energía (INQUIMAE), Facultad de Ciencias Exactas y Naturales, Universidad de Buenos Aires, Buenos Aires, Argentina

M. G. Lagorio (ed.), Comprehensive Series in Photochemical and Photobiological Sciences, Fluorescence of Living Plants 1, https://doi.org/10.1007/17361_2024_1
Published online: 25 March 2025

luminescence resulting from the electronic transition between states with the same spin multiplicity, which proceeds after excitation by energy absorption. During this process, the fluorophore emits part of the absorbed energy as electromagnetic radiation of lower energy (longer wavelength) [2].

The earliest documented examples of fluorescence in the literature are associated with natural substances employed by Native Americans for medicinal purposes. Exploring this historical overview is genuinely interesting and engaging [3].

In 1492, Christopher Columbus reached the Antilles from Spain, marking the onset of the conquest of America. This encounter between two vastly distinct cultures ultimately led to the Spanish colonization of a portion of the region. From the European viewpoint, this arrival was celebrated as the "Discovery of America", whereas, from the standpoint of indigenous peoples, it was regarded as a European invasion. Regardless of one's interpretation of this complex process, it undeniably left an indelible mark on the course of history.

For Europeans, this event signified more than just the chance to conquer lands and extract precious minerals; it also presented an opportunity to acquire novel insights into previously unknown animal and plant species, with many of the latter offering applications in the field of natural medicine (Fig. 1).

In fact, the first report of fluorescence dates back to the sixteenth century AD and is related to the emission of light from natural products being brought from America to Europe. The premiere documentation of a fluorescence process emerges from the

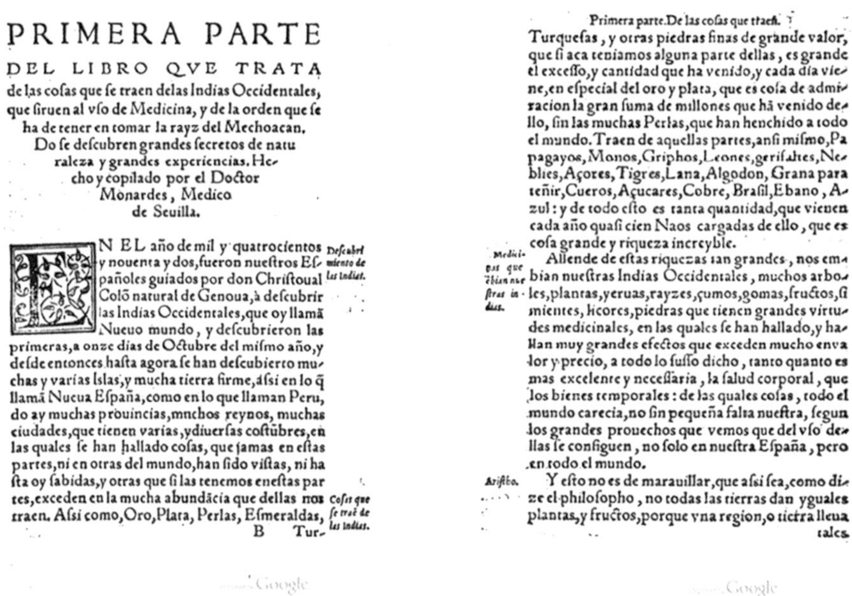

PRIMERA PARTE
DEL LIBRO QVE TRATA
de las coſas que ſe traen delas Indias Occidentales, que ſiruen al vſo de Medicina, y de la orden que ſe ha de tener en tomar la rayz del Mechoacan. Do ſe deſcubren grandes ſecretos de natu raleza y grandes experiencias. Hecho y copilado por el Doctor Monardes, Medico de Seuilla.

EN EL año de mil y quatrocientos y nouenta y dos, fueron nueſtros Eſpañoles guiados por don Chriſtoual Colõ natural de Genoua, à deſcubrir las Indias Occidentales, que oy llamã Nueuo mundo, y deſcubrieron las primeras, a onze dias de Octubre del miſmo año, y deſde entonces haſta agora ſe han deſcubierto muchas y varias Islas, y mucha tierra firme, aſsi en lo q̃ llamã Nueua Eſpaña, como en lo que llaman Peru, do ay muchas prouincias, muchos reynos, muchas ciudades, que tienen varias, y diuerſas coſtũbres, en las quales ſe han hallado coſas, que jamas en eſtas partes, ni en otras del mundo, han ſido viſtas, ni haſta oy ſabidas, y otras que ſi las tenemos eneſtas partes, exceden en la mucha abundãcia que dellas nos traen. Aſsi como, Oro, Plata, Perlas, Eſmeraldas, Tur-

Deſcubrimiento de las Indias.

Coſas que ſe traẽ de las Indias.

B

Primera parte. De las coſas que traen. 1

Turqueſas, y otras piedras finas de grande valor, que ſi aca teniamos alguna parte dellas, es grande el exceſſo, y cantidad que ha venido, y cada dia viene, en eſpecial del oro y plata, que es coſa de admiracion la gran ſuma de millones que hã venido dello, ſin las muchas Perlas, que han henchido a todo el mundo. Traen de aquellas partes, anſi miſmo, Papagayos, Monos, Griphos, Leones, geriſaltes, Neblies, Açores, Tigres, Lana, Algodon, Grana para teñir, Cueros, Açucares, Cobre, Braſil, Ebano, Azul: y de todo eſto es tanta quantidad, que vienen cada año quaſi cien Naos cargadas de ello, que es coſa grande y riqueza increyble.

Allende de eſtas riquezas tan grandes, nos embian nueſtras Indias Occidentales, muchos arboles, plantas, yeruas, rayzes, çumos, gomas, fructos, ſimientes, licores, piedras que tienen grandes virtudes medicinales, en las quales ſe han hallado, y hallan muy grandes efectos que exceden mucho en valor y precio, a todo lo ſuſſo dicho, tanto quanto es mas excelente y neceſſaria, la ſalud corporal, que los bienes temporales: de las quales coſas, todo el mundo carecia, no ſin pequeña falta nueſtra, ſegun los grandes prouechos que vemos que del vſo dellas ſe conſiguen, no ſolo en nueſtra Eſpaña, pero en todo el mundo.

Medicinas que ẽbian nueſtras indias.

Y eſto no es de marauillar, que aſsi ſea, como dize el philoſopho, no todas las tierras dan yguales plantas, y fructos, porque vna region, o tierra lleua tales.

Ariſtot.

Fig. 1 Some introductory fragments from the book authored by Nicolás Monardes: "*Historia medicinal de las cosas que se traen de nuestras Indias Occidentales que sirven en Medicina, Primera parte*" (1565) [4]

pages of a book authored by the physician and botanist, Nicolás Monardes. The introductory fragments of this book are shown in Fig. 1 [4].

FIRST PART
OF THE BOOK DEALING WITH
things brought from the West Indies, which serve for the use of Medicine and the order to be followed to use the root of Mechoacan. Where great secrets of nature and vast experiences are revealed. Compiled and written by Dr. Monardes, a physician from Seville. Translation of the text in old Spanish from Fig. 1.

In the year 1492, our Spaniards were led by Don Cristobal Colon, a native of Genoa, to discover the West Indies, which we now call the New World. They discovered them on the eleventh day of October of the same year, and since then, many various islands and much mainland have been discovered, both in what is called New Spain and in what they call Peru, where there are many provinces, many kingdoms, and many cities that have various and diverse customs. In these places, things have been found that have never been seen or known in these parts or in other parts of the world, and others that, if we have them in these parts, surpass them in the great abundance brought to us from there. Just as with Gold, Silver, Pearls, Emeralds, Turquoises, and other precious stones of great value, if we already had some of them here, the excess that has arrived, and continues to arrive every day, is substantial. In particular, concerning Gold and Silver, it is truly admirable the vast sum of millions that has come from there, not to mention the numerous Pearls that have enriched the entire world. They also bring from those parts, parrots, monkeys, griffin birds, lions, gerfalcons, falcons, goshawks, tigers, wool, cotton, cochineal for dyeing leather, sugars, copper, brazilwood, ebony, indigo: and there is such a quantity of all these that nearly a hundred ships laden with them arrive every year, which is a great thing and an incredible wealth. Beyond these immense riches, our West Indies send us many trees, plants, herbs, roots, juices, gums, fruits, seeds, liquors, and stones that possess significant medicinal values. In them, very powerful effects have been and continue to be discovered, which far exceed in value and price all the aforementioned, considering how much more excellent and necessary bodily health is than temporal goods.

Nicolás Bautista Monardes Alfaro hailed from a family of Genoese origin (Monardis) that settled in Seville in the fifteenth century. His exact birth date remains uncertain, ranging between 1493 and 1508 [5], depending on various sources. He passed away in 1588. Monardes held a Bachelor of Medicine (University of Alcala, 1533) and earned a doctorate in 1547 (University of Seville). He authored "Historia Medicinal de las cosas que se traen de nuestras Indias Occidentales" (published in three parts in 1565, 1571, 1574) (Fig. 2).

Monardes held a distinct fascination for natural products hailing from America, often referred to as "the New World", with a specific focus on their potential applications in the field of medicine. Interestingly, he never travelled to America. He traded natural products that reached him through intermediaries from those distant lands. He set up a botanical garden at his home where he planted and studied American plants (Fig. 3) [6, 7].

Fig. 2 Monardes portrait. Image of public domain included in his book *Historia medicinal de las cosas que se traen de nuestras Indias Occidentales que sirven en Medicina*. The portrait was published with a caption that reads: "*Elogio hecho por el Illust. S. Gonçalo Çati eco de Molina, Retrato del autor que se vee en su Museo*" Eulogy made by the illustrious Senior Gonçalo Çati, eco de Molina, Portrait of the author which can be seen in his Museum

Fig. 3 Commemorative plaque of Monardes Medicinal Botanical Garden in Seville. The plaque reads (translated to English): "*At this site stood the Medicinal Botanical Garden of Nicolás Bautista Monardes Alfaro (1493–1588), a universal figure from Seville who introduced American medicinal knowledge to Europe. In honor of the fourth centenary of his passing, the esteemed municipality chose to install this ceramic plaque*"

In his book, Monardes documented an intriguing blue hue (though he was unaware it was actually blue fluorescence) observed in infusions of a specific variety of wood sourced from Mexico, called *Lignum nephriticum* (wood related to the kidneys) [8, 9]:

Fig. 4 Species *Eysenhardtia polystachya.* The Aztecs utilized the wood from this tree to craft infusions with a distinctive blue "opalescence", known for its therapeutic effects on the urinary system. [From Marrovi—Trabajo propio, CC BY-SA 2.5 mx, https://commons.wikimedia.org/w/index.php?curid=16135697. (This image is licensed under the Creative Commons Attribution-Share Alike 2.5 Mexico license)]

"*Toman el palo y hacen unas tajaditas muy delgadas y no muy grandes y échanlas en agua…dentro de media hora se comienza el agua a poner con un color azul muy claro y cuánto más más azul se torna…*" that may be translated as:

> They extract thin and relatively small slices from the wood, submerging them in water. Approximately thirty minutes later, the water starts adopting a subtle pale blue hue, intensifying with prolonged exposure.

Monardes harnessed the therapeutic potential of this infusion in addressing kidney and urinary ailments. The Aztecs had long been acquainted with and utilized the advantageous properties of this tea derived from the wood of the *Eysenhardtia polystachya* species [10] (Fig. 4).

Monardes' book not only marks the first documentation of fluorescence but also introduces its application as a quality control criterion. Specifically, the text emphasizes:

> *El palo de la orina ha de hacer el agua "azul" y advierta que este palo que llamo de la orina y ijada, haga el agua azul, porque si no la hace azul no es lo verdadero porque traen ahora un palo que hace al agua amarilla y éste no es el que aprovecha, sino que haga el agua azul porque el tal que la hiciere azul será el verdadero*

which means:

> The urine stick must tint the water "blue" and take note that the stick, referred to as the urine and spleen stick, should indeed turn the water blue. Failure to do so signifies its lack of authenticity, as there is now a stick in circulation that turns the water yellow—yet, this is not the beneficial one. The true indicator lies in turning the water blue, affirming that the authentic stick is the one capable of such a transformation.

Fig. 5 Franciscan friar Bernardino de Sahagún (original name: Bernardo de Rivera o Riveira) [12]

Additional crucial documentation concerning these initial records of the "urine stick" can be found in the work entitled *Historia general de las cosas de la Nueva España; Codice Florentino* (General History of the Things of New Spain: The Florentine Codex[1]), authored by the Franciscan Friar Bernardino de Sahagún [11]. In this book, the urine stick is referred as "coatli" (meaning water snake in Náhuatl[2]) (Fig. 5).

Bernardino de Sahagún (1499–1590), who was born in Spain and passed away in Mexico, wrote this book in 12 volumes in both Náhuatl and Spanish [12]. This important and extensive work, on the life of native inhabitants of central Mexico, was assigned a preservation location: the *Biblioteca Medicea Laurenciana* in Florence, Italy; hence its name, Florentine Codex (Fig. 6).

In 1646, Athanasii Kircheri, a German Jesuit residing in Rome, provided a comprehensive discussion on the hues he perceived while preparing infusions of *Lignum nephriticum* in his work "*Ars Magna Lucis et Umbræ*" (The Great Art of Light and Shadow) [13]. He made a clear distinction between the yellow tint of "transmitted light" through the solution and the blue hue, which he identified as "reflected light" and which would in the future be recognized as fluorescence emission.

More than 400 years after it was first observed, Acuña et al. reported that the fluorophore responsible for the blue fluorescence of the infusions of *Lignum nephriticum* was a four-ring tetrahydromethanobenzofuro[2,3-d] oxacine (called matlaline) (Fig. 7) [14]. These authors also found that matlaline was not present in

[1]Codex: ancient manuscript book

[2]Náhuatl refers to a group of closely related indigenous languages spoken by the Nahua people in Mexico and parts of Central America.

Fig. 6 General History of the Things of New Spain: The Florentine Codex, Vol. 3, fo 141 (Left image) [Image 597 of Volume 3 | Library of Congress (loc.gov)]. Enlarged image depicting an indigenous person cutting wood ("*Coatli*" in the Náhuatl Language) from Species *Eysenhardtia polystachya* (Top Right); description of the image in Náhuatl (Center Right); Text in Spanish (Bottom Left). The text means: "There is a wild tree or bush, which is called *coatli* from which they make… . It is very ductile and if you put it in water, it makes the water of blue color and this water is medicinal for urine". (These books were written between 1540 and 1585 AD. https://www.loc.gov/resource/gdcwdl.wdl_10096_003/?sp=597&r=-0.218,0.327,1.513,0.959,0. Accessed on April 2024)

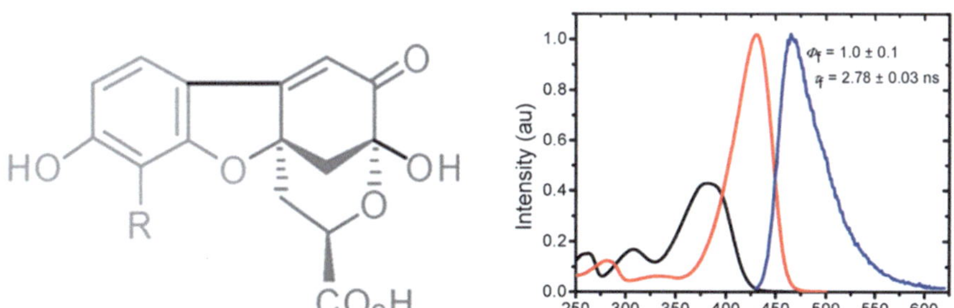

Fig. 7 Matlaline molecular structure and electronic spectra (red line: absorption spectrum at pH = 9, blue line: fluorescence spectrum (pH = 9) and black line: absorption spectrum at pH = 4. At pH = 4, the species does not emit. (Figures reprinted (adapted) with permission from [14, p. 3020]. Copyright 2009 American Chemical Society)

Fig. 8 David Brewster portrait. (From Unknown—http://www.stereoskopie-shop.com/stereoskope/stereoskope-001.htm, Public domain, https://commons.wikimedia.org/w/index.php?curid=16528842)

the plant but was the end product of spontaneous oxidation of some of the tree's flavonoids [14].

In 1833, David Brewster, a Scottish scientist (1781–1868), made the pioneering observation of **chlorophyll fluorescence** [15]. However, he did not characterize it as fluorescence; rather, he simply referred to it as a beam of light. Detailing the process, he observed that when sunlight passed through an alcohol solution of leaves, a red beam became discernible: "In making a strong beam of the sun's light pass through the green fluid, I was surprised to observe that its colour was a brilliant red complementary to the green".

Remarkably, he also noticed that: "By making the ray pass through greater thickness in succession, it became first orange and then yellow and yellowish-green…" firstly, reporting on fluorescence inner filter (also called light re-absorption) processes in concentrated solutions (Fig. 8).

This level of detailed observation shows this scientist's ability to detect with his eyes a phenomenon that would be described in depth several years later. In fact, the quantification of light re-absorption processes is a subject that is still being studied and modelled today (see Chapter "From Space to Cells: Physical Models for Light Re-absorption and Scattering").

David Brewster was licensed as a clergyman in 1800, at the University of Edinburgh. However, despite his theological and humanistic formation, he was attracted to the natural sciences and became an extremely distinguished scientist in the field of optical physics. He studied the polarization of light and the birefringence of crystals under compression. He also discovered photoelasticity. He developed the

Fig. 9 From George G. Stokes—Memoir and Scientific Correspondence of the Late Sir George Gabriel Stokes, Bart, Public domain, https://commons.wikimedia.org/w/index.php?curid=3127353

kaleidoscope and improved the stereoscope. A complete description of the life of this prestigious man was written by his daughter [16].

By these years, the phenomenon of fluorescence was observed for various materials but was described using non-specific terms as "reflectance" or "dispersion". In 1845, Sir John Herschel observed fluorescence from solutions of quinine sulphate but he referred to it as "epiploic dispersion". The term "epiploic" was constructed to describe surface colour and was inspired by the greek word "epipole" meaning surface [17].

It was not until 1852 that Sir George Stokes (1819–1903) coined the term "fluorescence" to describe the phenomena wherein specific materials absorbed short-wavelength light, like UV light, and subsequently emitted light at a longer wavelength, frequently within the visible spectrum. The chosen name was inspired by the mineral fluorite, renowned for its distinctive emission of visible light when exposed to UV illumination [18] (Fig. 9).

In a German-language article entitled "Beziehungen zwischen assimilation, absorption und fluoreszenz im chlorophyll des lebenden blattes" (Relationships between assimilation, absorption and fluorescence in the chlorophyll of the living leaf), Müller proposed a potential correlation between the in vivo fluorescence of chlorophyll and the process of photosynthesis [19]. This study introduced the exciting prospect of deducing photosynthetic information from light emission signals.

In 1931, Professor Hans W. Kautsky and his collaborator A. Hirsch conducted groundbreaking research on chlorophyll fluorescence and carbon dioxide assimilation. When they exposed dark-adapted leaves to light, they observed a variation in chlorophyll fluorescence as a function of time [20, 21]. Their observations were published in a concise one-page article (Kurze 0riginalmitteilungen: original short communication) entitled "Neue Versuche zur Kohlensäureassimilation, New Experiments on Carbon Dioxide Assimilation" in the journal "Naturwissenschaften" [20]. They observed an initial variation in fluorescence which was independent of both temperature and the presence of cyanide, which acted as a poison. Therefore, the authors inferred that the fast rise to a maximum intensity was related to the primary photochemical processes of photosynthesis. After reaching a maximum, the fluorescence began to decay in a slow process which displayed an inverse correlation with the experimental curve of CO_2 uptake published by Otto Warburg in 1920 [22]. In their work, they stated: "the experiments confirmed the hypothesis that the higher the proportion of absorbed radiation converted into thermal energy, the lower the brightness of chlorophyll fluorescence". They also discussed the excitation source that gave rise to chlorophyll fluorescence, stating that "Ultraviolet light (quartz analysis lamp) and visible blue light are equally effective in producing the phenomena, provided their intensity ratios are reasonably comparable". This short article would be a turning point in the use of chlorophyll fluorescence as a tool for inferring photosynthetic parameters and plant health.

Hans Kautsky (1891–1966) was an Austrian chemist. In his youth he studied painting but then turned to the sciences, studying chemistry at the Technical University of Charlottenburg from 1911 to 1917. He received his doctorate in 1922, which was carried out under the direction of Herbert Freundlich, at the Kaiser Wilhelm Institute, in Germany. It is curious to note that the subject of his doctoral thesis on unsaturated silicon compounds was far from the study of photosynthetic systems that would make his name in the field of photobiology (Fig. 10).

To diverge momentarily from this historical exposition and offer a personal opinion, I must confess a desire for more detailed experimental information regarding the "observation" of fluorescence intensity in Kautsky and Hirsch's article. While the literature indicates direct visual observation, it is common knowledge among practitioners in chlorophyll fluorescence that such observation is not straightforward in intact leaves. This experimental challenge in Kautsky and Hirsch's study adds significant depth to their already valuable contribution.

For those, like myself, who ponder the feasibility of visually observing chlorophyll fluorescence in leaves and its modulation by light intensity, there exists an enlightening video accessible on the Internet (Chlorophyll fluorescence: The glow of plants (youtube.com). This video offers a lucid and didactic demonstration of the process.

Fig. 10 Hans Kautsky. (From https://ru.wikipedia.org/w/index.php?curid=6293282. Authorship: unknown. http://research.uni-leipzig.de/agintern/uni600/ug289.htm, Fair Use)

In 1940, an article by McAlister and Myers explored relations between chlorophyll fluorescence and carbon dioxide uptake, in wheat and in *Chlorella* under different conditions [23]. The observations from numerous induction curves, acquired across various conditions, could be delineated by two processes. One process entailed an inverse correlation between the rate of CO_2 uptake and fluorescence intensity, while the other exhibited a direct correlation. So, they found that the fast process of chlorophyll fluorescence change was far simpler than the complex uptake of carbon dioxide. Jack Myers (1913–2006) was an American biologist, professor at Texas University with expertise in photosynthesis and phototropism. His research made relevant contributions on light utilization in photosynthesis by algae. A biography of this distinguished researcher was written by Elisabeth Gantt [24].

In the 1950s, significant contributions by scientist Duysen began to illuminate our understanding of photosynthesis. Louis Nico Marie Duijsens (Duysens) (1921–2015) was born in Heerlen, Limburg. He studied mathematics and physics in Utrecht, where he received his doctoral degree in 1947. Around this time, he changed the spelling of his family name from Duijsens to Duysens, the name by which he became scientifically known. In 1946, Duysens was appointed a staff member of the Biophysics Group of the University of Utrecht and the Delft University of Technology [25].

It is fascinating to note that his first investigations into chlorophyll fluorescence on cyanobacteria provided the initial clues regarding the existence of two distinct photosystems. Remarkably, the observation that fluorescence induced by phycobilins exhibited a higher quantum yield compared to direct excitation of chlorophyll suggested the presence of two photosystems: one exhibiting strong fluorescence selectively excited via phycobilins, and another displaying weaker fluorescence excited via chlorophyll [26, 27].

Approximately a decade later, Duysen et al. utilized absorption difference spectroscopy and manipulation of excitation wavelengths to go further ahead. When they excited the less fluorescent chlorophyll pool (Photosystem I) at 690 nm (far-red) cytochrome underwent oxidation, whereas when they excited via phycobilin the fluorescent chlorophyll pool (Photosystem II), they observed that cytochrome experienced reduction and water was oxidized with oxygen release. These findings led to the accurate deduction that light reactions in photosynthesis were driven by two different photosystems that operated consecutively [28].

In 1963, Duysens and Sweers assembled an instrument for chlorophyll fluorescence analysis, laying the foundation for what is now recognized as the Pulse Amplitude Modulated (PAM) Chlorophyll Fluorometer. Another crucial finding of these authors revealed that the fluorescence output of all oxygenic photosynthetic systems hinged on the redox status of a quencher known as Q, which was later identified as Quinone A [29].

Several years later from fluorescence measurements conducted on *Chlorella* Bonaventura and Myers [30] collected experimental evidence findings for a regulatory mechanism that governed the distribution of excitation energy between the two photosystems (I and II).

Although it has been verified that both photosystem II (PSII) and photosystem I (PSI) contribute to the initial chlorophyll fluorescence (F_0) at room temperature, it is usually considered that only PSII contributes to the variable fluorescence [21]. Nevertheless, in 2021, Schreiber and C. Klughammer published evidence for variable chlorophyll fluorescence of photosystem I [31]. In the literature, it is typically discussed that any variable fluorescence from PSI is negligible and can be disregarded [32]. However, it is a point that can still be considered under discussion.

In the context of the study of variable chlorophyll fluorescence, another important scientist, Reto Strasser, should be remembered for his significative contributions. He was a pioneer in researching the fast phase of chlorophyll-a fluorescence, known as JIP or OJIP test. His initial work related to OJIP curve was published in the form of "Yellow Books"[3] between 1967 and 2010 [33–35].

In the 1980s and 1990s relationships between the quantum yield of photosynthetic electron transport and the quenching of chlorophyll fluorescence were published [36–38]. Krause and Weis were pioneers in describing the non-photochemical quenching which represents the heat dissipated by the plant, a protection mechanism that is activated when there is excess in absorbed light.

[3] Yellow Books: The volume 1 of the Yellow Book Series contains all of the publications made by Professor Reto J. Strasser and collaborators for the period from 1967 until 1979. University of Geneva.

Methods of chlorophyll fluorescence analysis that utilize artificial light sources for excitation are referred to as active methods. In the 1970s, the initial endeavour to quantify fluorescence passively—without any artificial excitation source—marked a pivotal moment, sparking intensified research into Sun-Induced Fluorescence (SIF) [39, 40].

Over time, there has been a notable evolution in the analysis of chlorophyll fluorescence, encompassing advancements not only in its theoretical interpretation but also in the scales of field detection. In 1957, the dawn of the space age was marked by the launch of the first artificial satellite, the Sputnik I. From this point in history, the acquisition of information at a distance, without contact with the sample, would become each day more captivating.

In this context, remote sensing of chlorophyll fluorescence from satellites appeared as an attractive methodology in environmental monitoring and research. It provides a comprehensive and large-scale perspective, allowing for the monitoring of chlorophyll fluorescence over vast areas that would be impractical or impossible to cover through traditional ground-based methods. This capability enables researchers to study spatial and temporal patterns of photosynthetic activity across diverse ecosystems, from terrestrial vegetation to aquatic environments. Additionally, remote sensing from satellites offers frequent and consistent data acquisition, providing a continuous stream of information over time. This temporal resolution allows for the detection of seasonal variations, responses to environmental changes, and long-term trends in chlorophyll fluorescence, aiding in the understanding of ecosystem dynamics and responses to climate change. Furthermore, satellite-based remote sensing is non-invasive and minimizes disturbance to natural habitats, making it an environmentally friendly option for monitoring chlorophyll fluorescence.

Several space missions involving chlorophyll fluorescence have been conducted in recent years and are also planned for the near future. The first observation of SIF from space was performed on the Fourier Transform Spectrometer TANSO-FTS (EORC—JAXA) onboard the Japan's Greenhouse gases Observing SATellite (GOSAT) which was launched on January 23, 2009 [41, 42]. Other missions were: The "Orbiting Carbon Observatory-2 (OCO-2)" [43], the Chinese Carbon Dioxide Observation Satellite Mission (TanSat) [44], the Greenhouse gases Observing Satellite-2 (GOSAT-2) [45], the Orbiting Carbon Observatory-3 (OCO-3) [46], the Global Ozone Monitoring Experiment–2 (GOME-2) [47], the Scanning Imaging Absorption Spectrometer for Atmospheric Chartography (SCIAMACHY) [48], and the Tropospheric Monitoring Instrument (TROPOMI) [49] can obtain information of chlorophyll fluorescence from space. Specifically, for remote solar-induced fluorescence measurements, the European Space Agency (EA) has scheduled the launch of the "Fluorescence Explorer" (FLEX), for the year 2025 [50, 51].

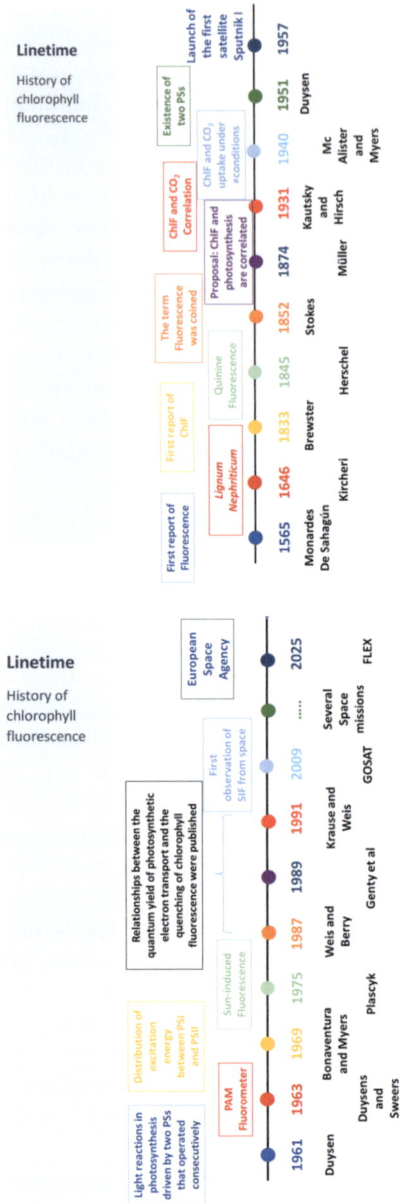

Video: Chlorophyllfluorescence: The glow of plants (youtube.com). How to make the fluorescence of plants and the Kautksy Effect visible to the human eye, with super easy tools. A Film by Institut für Bio- und Geowissenschaften, Pflanzenwissenschaften (IBG-2), Forschungszentrum Jülich). Accessed on February 2025.

References

1. Braslavsky SE (2007) Glossary of terms used in photochemistry, 3rd edition (IUPAC recommendations 2006). Pure Appl Chem 79:293
2. Lagorio MG, Cordon GB, Iriel A (2015) Reviewing the relevance of fluorescence in biological Systems. Photochem Photobiol Sci 14:1538–1559
3. Valeur B, Berberan-Santos MN (2011) A brief history of fluorescence and phosphorescence before the emergence of quantum theory. J Chem Educ 88:731
4. Monardes N (1565) in Historia medicinal de las cosas que se traen de nuestras Indias Occidentales que sirven en Medicina, p 1
5. Acuña AU, Amat-Guerri F (2008) Early history of solution fluorescence: the *Lignum nephriticum* of Nicolás Monardes. In: Berberan-Santos MN (ed) Fluorescence of supermolecules, polymers and nanosystems. Springer series on fluorescence, vol 4. Springer, Berlin, p 3
6. Guerra F (1961) Nicolás Bautista Monardes: su vida y obra, ca. 1493–1588. D.F. Compañía Fundidora de Fierro y Acero de Monterrey, 225 páginas
7. Boxer CR (1963) Two pioneers of tropical medicine: Garcia d'Orta and Nicolás Monardes. Wellcome Historical MedicalLibrary, London. Two pioneers of tropical medicine: Garcia d'Orta and Nicolás Monardes [electronic resource]: Boxer, C. R.(Charles Ralph), 1904– : Free Download, Borrow, and Streaming: Internet Archive. https://ia903208.us.archive.org/16/items/b20457248/b20457248.pdf. Accessed on Feb 2025
8. Safford WE (1916) The botanical identity of lignum nephriticum. Science 43:432–433
9. Müller NJC (1874) Beziehungen zwischen assimilation, absorption und fluoreszenz im chlorophyll des lebenden blattes. Jahrbücher für Wissenchaftliche Botanik 9:42–49
10. Acuña AU (2007) More thoughts on the narra tree fluorescence. J Chem Educ 84:231
11. de Sahagún B (1560–1564) In: de Bustamante CM (ed) Florentine Codex, Historia General de las cosas de Nueva España, vol. III, pp 95 and 231
12. Fernández T, Tamaro E (2004) Biografia de Fray Bernardino de Sahagún. En: Biografías y Vidas. La enciclopedia biográficaen línea [Internet]. Barcelona, España. Available in https://www.biografiasyvidas.com/biografia/s/sahagun.htm. Accessed on Feb 2025
13. Kircheri A (1646) Ars Magna Lucis et Umbræ, ATHANASII KIRCHERI Fuldensis Buchonii e Soc. JESU Prasbyteri; ARS MAGNA LUCIS ET UMBRAE, In: X. Libros digesta. QUIBUS Admirandae Lucis [et] Umbrae in mundo, atque adeo universa natura, vires effectusque uti nova, ita varia novorum reconditiorumque speciminum exhibitoione, ad varios mortalium usus, panduntur – Google Play Libros. https://archive.org/details/athanasiikirche00kirc/page/414/mode/2up. Accessed on Feb 2025
14. Acuña AU, Amat-Guerri F, Morcillo P, Liras M, Rodriguez B (2009) Structure and formation of the fluorescent compound of *Lignum nephriticum*. Org Lett 11:3020–3023
15. Brewster D (1834) On the colours of natural bodies. Trans R Soc Edinb 12:538–545
16. Gordon Brewster MM (1870) The home life of Sir David Brewster. By his daughter. With a portrait. Edmonston & Douglas, 482 pages. Free access in the link: Afterwards Gordon Brewster (Margaret Maria) – Google Books.) https://archive.org/details/homelifesirdavi01gordgoog/page/2/mode/2up. Accessed on Feb 2025
17. Herschel JFW (1845) On a case of superficial colour presented by a homogeneous liquid internally colourless. Philos Trans R Soc Lond 135:143
18. Stokes GG (1852) On the change in the refrangibility of light. Philos Trans R Soc Lond 142: 463–562
19. Müller NJC (1874) Beziehungen zwischen assimilation, absorption und fluoreszenz im chlorophyll des 2793 lebenden blattes. Jahrbücher für Wissenchaftliche Botanik 9:42–49
20. Kautsky H, Hirsch A (1931) Neue Versuche zur Kohlensäureassimilation. Naturwissenschaften 19:964
21. Govindjee (1995) Sixty-three years since Kautsky: Chlorophyll a fluorescence. Aust J Plant Physiol 22:131–160

22. Warburg O (1920) Über die Geschwindigkeit der photochemischen Kohlensäurezersetzung in lebenden Zellen. II Biochem Z 103:188–217
23. McAlister ED, Myers J (1940) Time course of photosynthesis and fluorescence. Science 92: 241–243
24. Gantt E (2018) Myers JE 1913–2006. A biographical Memoir by Elisabeth Gantt. https://www.nasonline.org/publications/biographical-memoirs/memoir-pdfs/myers-jack.pdf. Accessed on Feb 2025
25. Govindjee, Pulles MP (2016) Louis Nico Marie Duysens (March 15, 1921–September 8, 2015): a leading biophysicist of the 20th century. Photosynth Res 128:223–234
26. Duysens LNM (1951) Transfer of light energy within the pigment systems present in photosynthesizing cells. Nature 168:548
27. Duysens LNM (1952) Transfer of excitation energy in photosynthesis. Doctoral thesis. State University Utrecht, Utrecht
28. Duysens LNM, Amesz J, Kamp BM (1961) Two photochemical systems in photosynthesis. Nature 190:510–511
29. Duysens LNM, Sweers HE (1963) Mechanism of two photochemical reactions in algae as studied by means of fluorescence. In: The Japanese Society of Plant Physiologists (ed) Studies on microalgae and photosynthetic bacteria; special issue of plant and cell physiology. University of Tokyo Press, Tokyo, pp 353–372
30. Bonaventura C, Myers J (1969) Fluorescence and oxygen evolution from *Chlorella pyrenodoisa*. Biochem Biophys Acta 189:366–383
31. Schreiber U, Klughammer C (2021) Evidence for variable chlorophyll fluorescence of photosystem I in vivo. Photosynth Res 149:213–231
32. Kalaji HM, Schansker G, Brestic M et al (2017) Frequently asked questions about chlorophyll fluorescence, the sequel. Photosynth Res 132:13–66
33. Tsimilli-Michael M (2020) Special issue in honour of Prof. Reto J. Strasser–Revisiting JIP-test: an educative review on concepts, assumptions, approximations, definitions and terminology. Photosynthetica 58(Special Issue):275–292. https://doi.org/10.32615/ps.2019.150
34. Kalaji HM, Goltsev V (2020) Special issue in honour of Prof. Reto J. Strasser—foreword. Photosynthetica 58(Special Issue). https://doi.org/10.32615/ps.2020.046
35. Erismann KH, Strasser RJ, Braendle R (1967) Untersuchungen zur photo-syntheseabhängigen Sulfidoxidation durch Algen und höhere Pflanzen, insbesondere durch Lemna minor. Verh Schweiz Naturforsch Ges 67:139–141
36. Genty B, Briantais J-M, Baker NR (1989) The relationship between the quantum yield of photosynthetic electron transport and quenching of chlorophyll fluorescence. Biochim Biophys Acta Gen Subj 990:87–92
37. Weis E, Berry J (1987) Quantum efficiency of Photosystem II in relation to 'energy'-dependent quenching of chlorophyll fluorescence. Biochim Biophys Acta 894:198–208
38. Krause GH, Weis E (1991) Chlorophyll Fluorescence and Photosynthesis: The Basics. Annu Rev Plant Biol 42:313–349
39. Plascyk JA (1975) The MK II Fraunhofer line discriminator (FLD-II) for airborne and orbital remote sensing of solar-stimulated luminescence. Opt Eng 14:144339
40. Plascyk JA, Gabriel FC (1975) The Fraunhofer line discriminator MKII—an airborne instrument for precise and standardized ecological luminescence measurement. IEEE Trans Instrum Meas 24:306–313
41. Guanter L, Frankenberg C, Dudhia A, Lewis PE, Gómez-Dans J, Kuze A, Suto H, Grainger RG (2012) Retrieval and global assessment of terrestrial chlorophyll fluorescence from GOSAT space measurements. Remote Sens Environ 121:236–251
42. Frankenberg C, Fisher JB, Worden J, Badgley G, Saatchi SS, Lee JE, Toon GC, Butz A, Jung M, Kuze A, Yokota T (2011) New global observations of the terrestrial carbon cycle from GOSAT: patterns of plant fluorescence with gross primary productivity. Geophys Res Lett 38: L17706. https://doi.org/10.1029/2011GL048738

43. Sun Y, Frankenberg C, Jung M, Joiner J, Guanter L, Köhler P, Magney T (2018) Overview of solar-induced chlorophyll fluorescence (SIF) from the orbiting carbon Observatory-2: retrieval, cross-mission comparison, and global monitoring for GPP. Remote Sens Environ 209:808–823
44. Du S, Liu L, Liu X, Chen J (2021) First investigation of the relationship between solar-induced chlorophyll fluorescence observed by TanSat and gross primary productivity. IEEE J Sel Top Appl Earth Obs Remote Sens 14:11892–11902
45. Imasu R, Matsunaga T, Nakajima M et al (2023) Greenhouse gases observing SATellite 2 (GOSAT-2): mission overview. Prog Earth Planet Sci 10:33
46. Eldering A, Taylor TE, O'Dell CW, Pavlick R (2019) The OCO-3 mission: measurement objectives and expected performance based on 1 year of simulated data. Atmos Meas Tech 12(4):2341–2370
47. Joiner J, Yoshida Y, Guanter L, Middleton EM (2016) New methods for the retrieval of chlorophyll red fluorescence from hyperspectral satellite instruments: simulations and application to GOME-2 and SCIAMACHY. Atmos Meas Tech 9:3939–3967
48. Burrows JP, Hölzle E, Goede APH, Visser H, Fricke W (1995) SCIAMACHY—scanning imaging absorption spectrometer for atmospheric chartography. Acta Astronaut 35:445–451
49. Zhao F, Ma W, Köhler P, Ma X, Sun H, Verhoef W, Zhao J, Huang Y, Li Z, Ratual AK (2022) Retrieval of red solar-induced chlorophyll fluorescence with TROPOMI on the Sentinel-5 precursor Mission. IEEE Trans Geosci Remote Sens 60:1–14
50. Coppo P, Taiti A, Pettinato L, Francois M, Taccola M, Drusch M (2017) Fluorescence imaging spectrometer (FLORIS) for ESA FLEX mission. Remote Sens 9:649
51. Drusch M, Moreno J, Del Bello U, Franco R, Goulas Y, Huth A, Kraft S, Middleton EM, Miglietta F, Mohammed G, Nedbal L, Rascher U, Schüttemeyer D, Verhoef W (2017) The FLuorescence EXplorer mission concept—ESA's earth Explorer 8. IEEE Trans Geosci Remote Sens 55:1273–1284

Chlorophyll Fluorescence and Photosynthesis

Pablo I. Calzadilla, Anja Krieger-Liszkay, and Giles N. Johnson

Contents

Abstract Chlorophyll fluorescence is a noninvasive method to estimate the photosynthetic performance of organisms, from cyanobacteria to plants. Since the first mentions of chlorophyll *a* fluorescence in the literature, almost a century ago, it has been successfully used to gain insights into the performance and regulation of

P. I. Calzadilla (✉)
Institute for Integrative Biology of the Cell (I2BC), Université Paris-Saclay, CEA, CNRS, Gif-sur-Yvette Cedex, France

Department of Earth and Environmental Sciences, Faculty of Science and Engineering, University of Manchester, Manchester, UK
e-mail: pablo.calzadilla@i2bc.paris-saclay.fr

A. Krieger-Liszkay
Institute for Integrative Biology of the Cell (I2BC), Université Paris-Saclay, CEA, CNRS, Gif-sur-Yvette Cedex, France
e-mail: anja.liszkay@i2bc.paris-saclay.fr

G. N. Johnson
Department of Earth and Environmental Sciences, Faculty of Science and Engineering, University of Manchester, Manchester, UK
e-mail: giles.johnson@manchester.ac.uk

M. G. Lagorio (ed.), Comprehensive Series in Photochemical and Photobiological Sciences, Fluorescence of Living Plants 1, https://doi.org/10.1007/17361_2025_11
Published online: 13 March 2025

photosynthesis under different physiological and environmental conditions. This was possible due to the development of a theoretical framework combined with simple and affordable instrumentation. Several approaches have been designed, ranging across acquisition timescales (femtoseconds to seasons) and spatial dimensions, allowing the study of photosynthesis from a molecular perspective up to an ecosystem level. Nevertheless, their correct implementation and data interpretation require an understanding of the conceptual basis and models proposed. In this chapter, we will review the theoretical framework of chlorophyll fluorescence and discuss the most commonly used approaches and related parameters in a plant physiology context. We highlight the significance of chlorophyll fluorescence for understanding the photosynthetic regulation of plants under fluctuating environments.

Keywords Electron transport, Fluorescence induction curves, Modulated chlorophyll fluorescence, Plant physiology, Plant stress responses

Photosynthesis is the metabolic process through which plants (and other oxygenic photosynthetic organisms) transform sunlight into chemical energy. Light energy is absorbed by pigments in their thylakoid membranes, with chlorophyll being the most important. When chlorophyll *a* absorbs light energy, this energy can be captured and used to drive chemical reactions (photochemistry), lost as heat, or reemitted as light at a longer wavelength – i.e., fluorescence. These different pathways exist in competition, such that an increase in one will result in a decrease in the others. Fluorescence is technically easy to measure with high resolution and provides a proxy to estimate photosynthesis and heat dissipation. Chlorophyll fluorescence measurements have been performed on samples ranging from isolated proteins up to whole ecosystems and on timescales from femtoseconds to seasons. In this chapter, we focus on the most widespread and accessible applications – measuring intact photosynthetic organisms (mostly plants) under normal physiological conditions on timescales from seconds to hours. We examine the theory underlying the analysis and discuss some issues and pitfalls associated with its application.

1 Chlorophyll Fluorescence in Photosynthesis Research

In photosynthetic organisms, light is harvested by two photosystems (photosystem I (PSI) and photosystem II (PSII)), located in a membrane system called the thylakoid. The photosystems use light energy to drive charge separation reactions, producing oxidizing and reducing molecules and triggering electron transport reactions in the thylakoid membranes. Electrons flow from water to PSII and from there to PSI through several electron carrier molecules, with concomitant translocation of H^+ to the thylakoid lumen (Fig. 1a). The resulting proton concentration gradient (ΔpH) and charge difference between the thylakoid lumen and the chloroplast stroma ($\Delta\psi$)

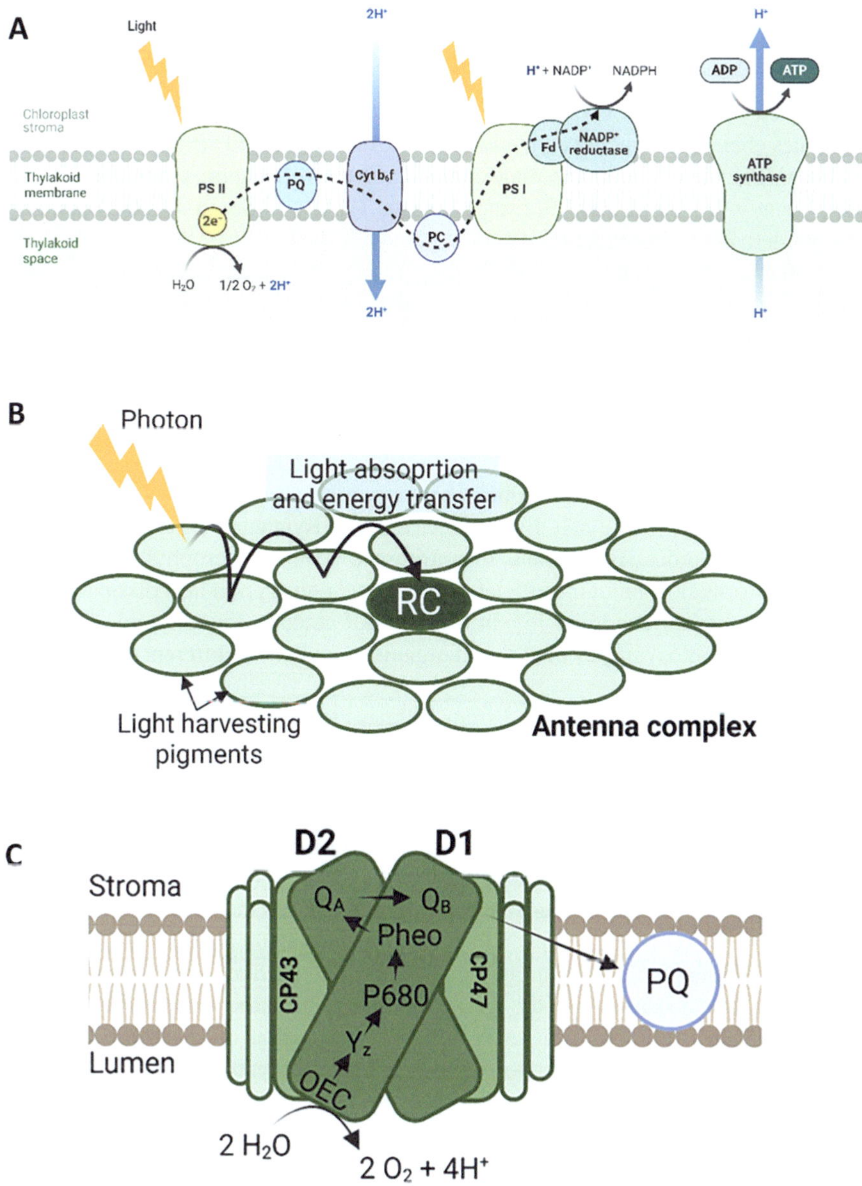

Fig. 1 (**a**) Photosynthetic linear electron flow (LEF). Electrons move from photosystem II (PSII) to photosystem I (PSI) in the thylakoid membranes with concomitant translocation of H^+ to the thylakoid lumen. The resulting proton concentration gradient (ΔpH) is used by a transmembrane ATP synthase to produce ATP. Beyond PSI, electron flows from ferredoxin (Fd) to the final electron acceptor $NADP^+$ via the ferredoxin-$NADP^+$ reductase (FNR), which generates chemical reducing equivalents in the form of NADPH. Both ATP and NADPH are used mainly by the Calvin-Benson-Bassham (CBB) cycle for sugar synthesis. (**b**) In plants and green algae, light absorption depends on pigments in light-harvesting complex (LHC) antenna (i.e., chlorophyll (Chl) *a* and *b*, carotenoids), which transfer the excitation energy ("excitons") to the reaction centers (RC) of the photosystems.

represent an energy store denoted as the proton motive force (*pmf*). A transmembrane ATP synthase uses this energy to produce ATP. Meanwhile, at PSI, the soluble electron donor ferredoxin (Fd) is reduced. Once reduced, electrons flow from Fd^- to the final electron acceptor $NADP^+$ via the ferredoxin-$NADP^+$ reductase (FNR). FNR generates chemical-reducing equivalents in the form of NADPH (Fig. 1a). Both ATP and NADPH generated in the photosynthetic electron transport chain (PETC) are used mainly by the Calvin-Benson-Bassham (CBB) cycle to fix CO_2-forming glyceraldehyde-3-phosphate (G3P) for sugar synthesis.

In plants and green algae, light absorption depends on pigments in light-harvesting complex (LHC) systems (i.e., chlorophyll (Chl) *a* and *b*, carotenoids), which transfer the excitation energy ("excitons") to the reaction centers of the photosystems (Fig. 1b). In the reaction centers, excited Chl *a* (P680 in PSII and P700 in PSI) triggers charge separation, which is the first photochemical reaction. Although this process is highly efficient, excited Chl *a* can return to its ground energy state by emitting fluorescence in the red/far-red spectrum [3]. The proportion of absorbed light energy used for photochemistry or dissipated non-photochemically (i.e., fluorescence, heat) can be defined as photosynthetic energy partitioning [4]. Since these processes compete, measuring the yield of chlorophyll fluorescence can give information about the efficiency of photochemistry and heat dissipation in a given physiological condition [5]. In the last half century, this principle has been used to study photosynthesis in different organisms and under different experimental conditions [6].

When a plant leaf is transferred from darkness to light, the chlorophyll fluorescence yield transiently increases, followed by a decrease in the second to minute time range (Fig. 2a). The fluorescence yield increase has been shown to reflect the reduction of electron acceptors in the thylakoid membrane [8]. Upon light absorption, excited P680 in the PSII reaction center undergoes charge separation and reduces a PSII-bound plastoquinone molecule termed Q_A (Fig. 1c). At the same time, $P680^+$ is re-reduced by an electron originating from the water-splitting apparatus. Reduced Q_A cannot accept another electron from P680 until it is oxidized by a second plastoquinone, Q_B. During this period, the reaction center is said to be "closed," since it cannot use the excitation energy absorbed by the antenna for electron transfer reactions. Therefore, a larger proportion of absorbed energy is released as fluorescence, leading to the observed increase in the yield of chlorophyll fluorescence (Fig. 2a) [2]. At room temperature, most fluorescence is emitted from

Fig. 1 (continued) Image based on [1]. (**c**) Upon light absorption, P680 in the PSII reaction center reduces a PSII-bound plastoquinone molecule termed Q_A via the redox intermediary pheophytin (Pheo). At the same time, $P680^+$ is re-reduced by an electron originating from the water-splitting apparatus (oxygen-evolving complex (OEC)) via a redox-active tyrosine residue (Yz). Reduced Q_A cannot accept another electron from P680 until it is oxidized by a second plastoquinone, Q_B. *CP43 and CP47* intrinsic antenna proteins of PSII reaction centre, *PQ* plastoquinone pool. (Imaged based on [2]. Created in Biorender.com)

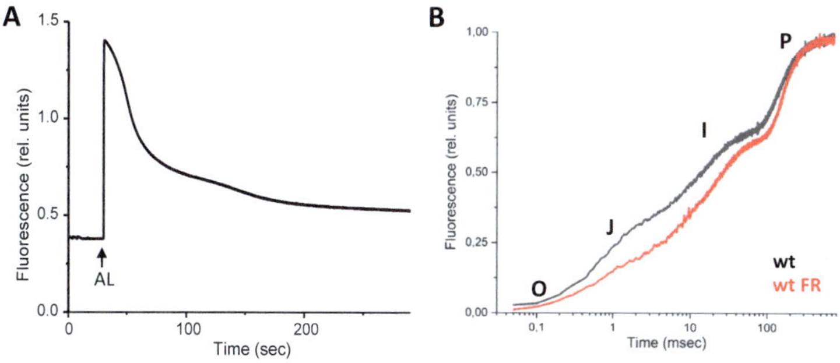

Fig. 2 (**a**) When a plant leaf is transferred from darkness to light (actinic light, AL), the chlorophyll fluorescence yield transiently increases, followed by its decrease in the second to minute range. This is known as the "Kautsky effect." Fluorescence trace observed when a leaf of *Arabidopsis thaliana* is exposed to 2,000 μmol m^{-2} s^{-1} of white light after 20 min of dark incubation. Measurement was performed with a PAM fluorometer. (**b**) Fast fluorescence increase observed when an *Arabidopsis thaliana* leaf is exposed to 1-s saturating light pulse (OJIP curve) (data kindly given by Maëlie Chazette, I2BC-Université Paris-Saclay, France). Black, wt leaf incubated in the dark for 20 min before measurement; Red, wt line incubated in the dark for 20 min, exposed to far-red light for 2 min before measurement. Based on the Q_A model [7], differences in the inflections J and I when the leaf is preincubated with far-red light (preferentially absorbed by PSI) are due to higher oxidation state of the electron carriers in the photosynthetic electron transport chain. Data were normalized to the maximum fluorescence level for comparison. Measurements were performed with a PAM fluorometer

light-harvesting antennae connected to PSII, and only a small and constant amount is emitted from the PSI antenna (about 10% at maximal Chl fluorescence level) [9].

The second phase of the curve, which is characterized by a slower decrease in fluorescence yield (called "fluorescence quenching," Fig. 2a), can be explained by a combination of processes. Electron transfer reactions via Q_B, PQ pool, cytochrome b_6f complex, plastocyanin, and oxidized P700 to ferredoxin are accelerated due to activation of downstream reactions (e.g., CBB cycle enzymatic activation, increasing CO_2 availability due to stomatal opening). Moreover, continuous illumination increases the probability of dissipation of absorbed energy as heat (non-photochemical quenching – NPQ) as well as other mechanisms of light responses such as state transitions [10, 11]. The increase and fall in the yield of chlorophyll fluorescence upon light illumination are denominated the "Kautsky effect" [6]. Methods to disentangle the different phases of transient fluorescence curves have been developed [12]. However, since the fluorescence yield will depend on the chlorophyll content, physiological status, and experimental conditions [13], measurements must be normalized to analyze photosynthetic energy partitioning.

Two different fluorescence levels are commonly used for normalization – the basal fluorescence of a dark-adapted leaf called F_o (all Q_A oxidized, PSII centers open) and the maximal fluorescence observed in the absence of photochemistry, F_m (all Q_A reduced, PSII centers closed) [7]. To estimate F_m, a state must be induced in which all reaction centers are closed. This can be achieved artificially by adding chemical inhibitors of electron transport such as DCMU (3-(3,4-dichlorophenyl)-

1,1-dimethylurea). DCMU binds to the Q_B-binding pocket and blocks electron transfer between Q_A and Q_B, leading to the closure of PSII reaction centers [14]. However, since the addition of chemicals impairs the physiological status of leaves, a different methodology, sometimes known as the "light doubling" method (or "saturating pulse" method), allows for F_m to be transiently reached without the addition of inhibitors [15, 16]. In this method, a short, high-intensity light flash is given. This induces all PSII to transiently close and, due to the short duration of the flash (typically 300 ms–1 s), photochemistry (beyond Q_A) and non-photochemical dissipation remain unchanged, and the maximum yield of fluorescence is reached. Comparisons of F_o and F_m under dark conditions and the fluorescence levels measured in the light – the steady-state level F_t (sometimes F_s or simply F) and the maximum in the light, F'_m, respectively – can give us insights into the photosynthetic process and photosynthetic energy partitioning. The methodologies used to determine these parameters will be addressed in the following sections.

2 Non-modulated Chlorophyll Fluorescence

At its simplest, fluorescence emission can be measured by shining a light on a leaf and measuring the light emitted at a longer wavelength. This is termed non-modulated fluorescence. Typical methodologies require dark adaptation of the samples (e.g., 20 min), which are then exposed to a short saturating light pulse (up to 1 s), and the kinetics of the fluorescence rise analyzed [8]. The changes occurring during the fast fluorescence increase (0.1–1,000 ms) have been related to the successive reduction of electron acceptors in the PETC, and they are known as the OJIP curve (where O is the origin, the minimum level; J and I are inflections; P is the peak, equivalent to F_m if measured in a saturating light) (Fig. 2b) [17, 18]. The theory assumes that PSII is the main contributor to variable Chl *a* fluorescence and that the fluorescence rise is related to PSII acceptor-side limitation (the rate of Q_A reduction) [8, 19]. This is known as the Q_A model and allows for estimating the maximum quantum yield of PSII and the electron transport along the PETC [7]. However, the exact mechanisms explaining these fluorescence phases are still a matter of debate (for a review, see [7]).

The increases from the basal fluorescence level "O" to the maximum level "P" are divided into the "photochemical" and "thermal" phases (O–J rise and J–I–P rise, respectively) [7, 20]. The J and I phases are often seen as a reflection of electron transport from Q_A^- to Q_B and to the full reduction of the PQ pool [21]. Meanwhile, it is known that PSI activity is reflected in the transition to the P phase [22]. In the presence of DCMU, the thermal phase is eliminated, and the maximum fluorescence level (P) is reached immediately after illumination (approx. 1 ms). Based on the Q_A model, the sigmoidal increase of chlorophyll fluorescence from the minimum O to the maximum P (without the inflections J and I) is due to the kinetics of reduction of Q_A [7]. In this regard, PSII reaction centers will exist in two states (i.e., quenched and unquenched, which will be related to the presence of Q_A or Q_A^-, respectively). The fluorescence rise will depend on the rate of reduction of Q_A, which is faster in the presence of DCMU. However, there are controversies concerning this explanation [7, 23].

When a single turnover flash is used as a light source, assumed to reduce Q_A in almost all PSII reaction centers, a maximum fluorescence level is reached, which is lower than the fluorescence level P [24, 25]. Additional flashes are needed to reach P, although charge separation is complete after the first flash. This discrepancy to the Q_A model can be explained by defining two energetically different equilibrium states of closed PSII centers: a charge-separated closed state of PSII (presence of Q_A^-, "$PSII_C$") and a closed PSII in the so-called light-adapted state (defined as "$PSII_L$"). $PSII_L$ will be responsible for the further increase in variable chlorophyll fluorescence observed under multiple turnover saturating flashes [26]. The transition from $PSII_C$ to $PSII_L$ upon illumination is related to conformational changes in the reaction center, triggered by electrostatic field changes and/or dielectric relaxation processes [27].

Although the interpretation of OJIP curves remains under discussion [7, 23], the Q_A model is useful when using chlorophyll fluorescence to monitor the physiological status of a given sample. The OJIP analysis has been used to study abiotic stress responses in model and non-model plant species [28, 29], in the screening of abiotic stress-tolerant varieties of crops [30, 31] as well as in ecological studies [32, 33]. Its wide implementation has been facilitated due to the development of relatively inexpensive and user-friendly fluorometers. However, due to the reasons previously stated, the interpretation and conclusions obtained with this methodology need to be considered cautiously.

3 Modulated Chlorophyll Fluorescence

Any light that falls on a leaf will, with a certain efficiency, result in the emission of fluorescence from chlorophyll. In chlorophyll fluorescence analysis, we want to know the efficiency of that emission. In simple non-modulated systems, as discussed above, a controlled light source is shone at a sample, and the fluorescence that results is measured, giving a signal proportional to the fluorescence yield. However, if any other light falls on the leaf, as is the case in any natural situation, it will also induce fluorescence emission. Thus, the signal we record will no longer correspond to the yield, limiting measurements to systems in which the light is controlled and constant. This issue was overcome with the use of modulated fluorometry systems and made widely available with the advent of commercially available pulse amplitude-modulated (PAM) fluorometers [34].

In PAM fluorometers, a fluorescence measuring light is modulated, meaning it is switched on and off with a certain frequency. The fluorescence detector is tuned to the same frequency, such that its output reflects only the fluorescence coming from the modulated light source and not from ambient or background light sources. The frequency and intensity of the measuring light are set to prevent it from inducing a significant net reduction of Q_A, so modulated fluorometers can be used to estimate the photosynthetic efficiency of leaves under a wide range of background illuminations, even direct sunlight [35].

The development of PAM fluorometers also massively increased the ease with which the saturation pulse method could be used, meaning measurements of F_m across different environmental conditions became straightforward [15, 16, 34]. This opened up the possibility of applying simple cheap measurements to estimate photosynthetic efficiency, even in the field. As a result, chlorophyll fluorescence analysis has become one of the most widely used and powerful techniques in plant physiology. Table 1 outlines the different parameters that can be obtained using chlorophyll fluorescence analysis, which will be described in the following sections.

3.1 Chlorophyll Fluorescence Analysis

A typical experiment with a PAM fluorometer starts with a dark-adapted leaf. Using non-stressed plants in lab conditions, this may only be for a short period (e.g., for 20 min), but often, and especially when plants are being stressed, a longer (overnight to 24 h) period can be used. This ensures that any processes regulating heat dissipation (see NPQ below) are inactivated and that the true F_m can be estimated. The fluorometer measuring light is then switched on to determine F_o, followed by a saturating flash, to determine F_m (Fig. 3). The difference between F_m and F_o is defined as variable fluorescence (F_v) and can be used to calculate the maximum quantum yield of PSII, known as F_v/F_m:

$$\frac{F_v}{F_m} = \frac{F_m - F_o}{F_m} \tag{1}$$

This estimates the maximum quantum yield efficiency of PSII in darkness, when all PSII centers are assumed to be "open" [36]. Then, a photosynthetically active light source is switched on (termed "actinic light") to drive photosynthesis, and saturating light pulses can be given at intervals to determine F'_m, the value of F_m in the light. The difference between F'_m and F_t (measured immediately before the flash) allows us to determine the quantum yield of PSII (ΦPSII or sometimes Y(II)) under the conditions being experienced, i.e., the proportion of light absorbed by PSII which is used in photochemistry [5, 36]:

$$\Phi \mathrm{PSII} = \frac{F'_m - F_t}{F'_m} \tag{2}$$

F_v/F_m and ΦPSII are routinely used to assess photosynthetic capacity in plants. The former is a "healthiness" indicator of the photosynthetic apparatus, with common

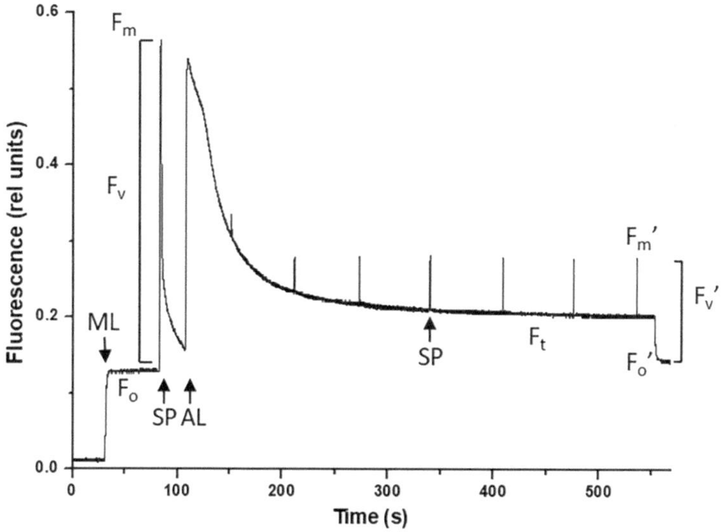

Fig. 3 Typical fluorescence trace obtained using a PAM fluorometer and *Arabidopsis thaliana* leaves. First, the fluorometer measuring light is switched on to determine F_o, followed by a saturating flash to determine F_m. Then, a photosynthetically active light (actinic light) source is switched on to drive photosynthesis. Saturating light pulses can be given at intervals to determine F'_m. The difference between F_m and F_o is defined as variable fluorescence (F_v) and can be used to calculate the maximum quantum yield of PSII, known as F_v/F_m. The difference between F'_m and F_t (the fluorescence level under illumination) allows us to determine the quantum yield of PSII (known as ΦPSII or sometimes Y(II), Table 1). Leaves were dark-adapted for 20 min and then illuminated with an actinic light of 500 μmol m^{-2} s^{-1}. The intensity of the saturating pulses was 4,500 μmol m^{-2} s^{-1}. *ML* measuring light, *AL* actinic light, *SP* saturating pulse, F_o basal fluorescence when measuring light is on, F_m maximal fluorescence in dark-adapted leaves, F_v variable fluorescence (F_m F_o); F'_m maximal fluorescence in light-adapted leaves, F_t fluorescence under continuous illumination, F'_o basal fluorescence levels reached after the actinic light is switched off

values around 0.8–0.85, depending on species, while lower values are associated with stress exposure and PSII photoinactivation (i.e., photoinhibition) [38]. Under known light conditions, ΦPSII can be used to estimate the relative PSII electron transport rates (ETR = ΦPSII multiplied by the light absorbed by the PSII antenna; Table 1). This has been widely adopted as a proxy measure for the photosynthetic activity of organisms [36, 39]; however, caution is needed when interpreting data. The amount of light incident on a leaf, which is captured by the PSII antenna, will vary depending on the chlorophyll content and the proportion of that chlorophyll associated with PSII (which can vary long-term due to photosynthetic acclimation and short-term due to state transitions). Different leaves will also have different optical properties. Leaf absorbance varies depending on leaf thickness and factors such as epicuticular wax pubescence (presence of hairs) and presence of non-photosynthetic pigments [40, 41], and an accurate estimate of absorbance requires the use of an integrating sphere.

Table 1 Commonly used fluorescence parameters, definition, abbreviations, and equations. Q light intensity (μmol m^{-2} s^{-1})

Parameter	Equation	Definition
F_v/F_m	$\frac{F_m - F_o}{F_m}$	Maximum quantum yield of PSII
ΦPSII (or Y (II))	$\frac{F'_m - F_t}{F'_m}$	Quantum yield of PSII
ETR	ΦPSII × Q × 0.84 × 0.5	PSII electron transport rates
NPQ	$\frac{F_m - F'_m}{F'_m}$	Non-photochemical quenching
F'_v/F'_m	$\frac{F'_m - F'_o}{F'_m}$	Maximum quantum yield of PSII under light conditions
qP	$\frac{F'_m - F_t}{F'_m - F'_o}$	Proportion of open PSII based on the "puddle" model [36]
qL	$\frac{F'_m - F_t}{F'_m - F'_o} \cdot \frac{F'_o}{F_t}$	Proportion of open PSII based on the "lake" model [37]
ΦNPQ	$\frac{F_t}{F'_m} - \frac{F_t}{F_m}$	Yield of regulated heat dissipation
ΦNO	$\frac{F_t}{F_m}$	Yield of nonregulated dissipation

It is common, including in the software associated with commercial machines, to assume a leaf absorbance of 0.84 and a partitioning to PSII of 0.5. These are, however, assumptions that are almost always wrong, to varying degrees. Acclimation processes modify photosynthetic complex content, including changing photosystem ratio and antenna size, with, for example, plants grown in shade light typically having more LHCII and less PSI [42–44]. Such changes will be reflected in measurements of chlorophyll *a*:*b* ratio since reaction centers contain only Chl *a*. This parameter indicates the amount of light-harvesting complexes relative to reaction centers and can vary substantially both between and within species depending on growth conditions [43–45]. Nevertheless, measurements of ETR can be useful to estimate photosynthetic activity, provided that the limitations are borne in mind. Additionally, the relationship between PSII turnover and carbon assimilation is not simple. This depends on environmental conditions and the availability of CO_2 relative to other electron acceptors [46, 47]. Variations in the relationship between ETR and photosynthesis, measured by gas exchange, can provide valuable information.

A third parameter commonly calculated using PAM fluorometers is non-photochemical quenching (NPQ). NPQ provides an estimate of the energy dissipated as heat in PSII and is calculated as the difference between F_m and F'_m over F'_m (Table 1) [48]:

$$\mathrm{NPQ} = \frac{F_m - F'_m}{F'_m} \tag{3}$$

NPQ represents the increase in heat dissipation relative to a reference level (the measured dark F_m) and as such only shows changes in heat dissipation. NPQ encompasses several fluorescence quenching processes with different induction/relaxation kinetics and relative contributions (e.g., state transitions, photoinhibition). However, the main contributor is a process termed high energy state quenching (qE), which plays a major role in photoprotection [49]. qE induction depends on the protonation of the PsbS protein of PSII and the formation of the xanthophyll zeaxanthin through de-epoxidation of violaxanthin in the xanthophyll cycle. Both processes are dependent on thylakoid lumen acidification, which triggers conformational changes in the PSII antenna [50]. Considering photosynthetic energy partitioning, photosynthesis and heat dissipation exist in competition, so qE induction (or high NPQ) will result in a lowering of ΦPSII. NPQ measurements can give insights into plant stress responses and acclimation [45, 51]. However, comparisons between treatments or species with different activity status of PSII (i.e., different F_v/F_m) should be taken cautiously since interpretations might be misleading. NPQ is discussed in more detail in Chap. 3.

4 Alternative Fluorescence Parameters

The three parameters described above are the workhorses of chlorophyll fluorescence analysis. Between them, they provide much of the key information about the photosynthetic apparatus – the maximum PSII efficiency (F_v/F_m, reflecting damage to PSII), the actual efficiency (ΦPSII), and the extent of photoprotective heat dissipation (NPQ). However, additional parameters have been developed that can be valuable in understanding photosynthetic responses in more detail.

4.1 *Photochemical Quenching and the PSII Redox State*

If we examine the processes that result in a loss of PSII efficiency, these are the extent to which heat dissipation competes with charge separation and the extent to which PSII is saturated in light. The former is reflected in changes in F_m, which can be measured either as a change in NPQ or through the parameter F'_v/F'_m (Table 1) The latter parameter has the advantage that it can be estimated in the absence of information about dark-adapted F_m, though it does require an accurate estimate of F'_o. F'_o corresponds to the basal fluorescence levels reached after the actinic light is switched off. Thus, F'_v/F'_m is equivalent to the dark F_v/F_m described above and provides a measure of the maximum quantum yield in the light, in the presence of photoprotective processes. In simple terms, the relationship between ΦPSII (actual quantum yield in the light) and F'_v/F'_m (maximum quantum yield if all PSII reaction centers were open) is determined by the proportion of open PSII, as defined by the parameter qP (Table 1) [36]:

$$qP = \frac{F'_m - F_t}{F'_m - F'_o} = \Phi PSII / \frac{F'_v}{F'_m} \tag{4}$$

This is an elegant and simple relationship, but unfortunately, it is somewhat simplistic. An underlying assumption in this model is that each PSII photosystem is isolated from all others (the so-called puddle model) [37]. In reality, energy absorbed by a photosystem with a closed reaction center can migrate to a different reaction center, increasing the probability that energy will be quenched photochemically for a given extent of saturation. This is defined as the "lake" model [37]. Kramer et al. [37] re-examined the models underlying fluorescence analysis and derived an alternative parameter to measure PSII open centers:

$$qL = \frac{F'_m - F_t}{F'_m - F'_o} \cdot \frac{F'_o}{F_t} = qP \cdot \frac{F'_o}{F_t} \tag{5}$$

It is important to be aware that these two models represent extreme cases, assuming complete connectivity or disconnection between reaction centers. In plants, the reality is probably closer to the lake model, with centers connected, but care should be taken not to overinterpret these parameters. Interestingly, ΦPSII is insensitive to reaction center connectivity models, so it is a more robust parameter. In addition to presenting the proportion of PSII reaction centers that are open, many authors prefer to show the proportion that is closed (1 − qP or 1 − qL). Depending on the data being shown, either can be useful.

4.2 *Alternative Parameters to Describe Non-photochemical Quenching*

A different but widely used approach to understanding the distribution between energy partitioning pathways has been proposed by Klughammer and Schreiber [52]. Broadly, the yield of fluorescence at any moment is a function of the competition with photochemistry and heat dissipation. In this analysis, heat dissipation is divided into regulated heat dissipation (i.e., induced NPQ) and nonregulated dissipation. All absorbed light is assumed to decay via either photochemistry or regulated or nonregulated dissipation. The *yield* of these three pathways therefore adds up to 1:

$$\Phi PSII + \Phi NPQ + \Phi NO = 1 \tag{6}$$

where ΦNO is the sum of nonregulated heat dissipation and fluorescence emission. The symbol Φ is often replaced by the Roman letter Y, representing yield, but the terms are identical. It is assumed that the rate constant for fluorescence (k_f) does not vary with changing conditions and is included in the nonregulated yield. As before, the calculations require a reference point, F_m, at which both photochemistry and regulated heat dissipation are absent. ΦPSII is as defined previously. The yields of the different dissipation pathways are defined as (Table 1)

$$\Phi\mathrm{NPQ} = \frac{F_{\mathrm{t}}}{F'_{\mathrm{m}}} - \frac{F_{\mathrm{t}}}{F_{\mathrm{m}}} \tag{7}$$

and

$$\Phi\mathrm{NO} = \frac{F_{\mathrm{t}}}{F_{\mathrm{m}}} \tag{8}$$

It is important to bear in mind that these terms describe the yields of each of the three defined pathways. So, if for example photosynthesis is blocked (e.g., by an inhibitor or by low temperature), the yield of the dissipative pathways must increase, even if the rate constants that define them are unchanged. If protective non-photochemical quenching is activated, at least one of the other two pathways will decrease in yield.

As with measurements of NPQ, the parameters ΦNPQ and ΦNO depend on the measurement of a reference dark level of fluorescence that includes any dissipative mechanisms that are present in the dark-adapted state. If a leaf has been exposed to a stress prior to the measurement and is therefore suffering from photoinhibition, this will result in a low F_{m} and therefore a higher ΦNO. If photoinhibition is induced during the measurement (i.e., after estimation of F_{m}), then heat dissipated by the damaged reaction centers will contribute to ΦNPQ, even though the underlying processes are identical. The different parameters described here are in the end simply different ways of presenting the same information. Depending on the situation, it may be preferable to use one or the other, but arguably, it does not add anything to show both in a single analysis.

4.3 *Deconvolution of NPQ*

The parameter NPQ quantifies processes that give rise to a reduction in F'_{m} relative to F_{m}, resulting from an increase in processes that dissipate energy as heat. As noted above, this is dominated by photoprotective high-energy-state quenching (qE; see Chap. 3). This quenching is rapidly induced upon exposure to excess light conditions and relaxes within a few minutes when the plant is returned to low light or darkness. In addition, other processes have been identified as contributing to NPQ, including state transitions, presence of xanthophyll zeaxanthin, and damage to the D1 protein of PSII (i.e., photoinhibition). These processes can be partially distinguished by performing measurements of quenching relaxation [53]. In such an analysis, a leaf is illuminated for a period, and F'_{m} is measured. The actinic light is then switched off, and quenching processes are allowed to relax in the dark. The recovery of F_{m} can then be followed by applying periodic saturating flashes. Care must be taken to ensure that these flashes do not themselves result in significant NPQ; however, provided that they are sufficiently spaced, it is possible to obtain an estimate of the relaxation of quenching. Approaches to deconvolute these decay kinetics have involved assuming two or more components, each approximating to an exponential decay [53, 54]. As a simple approach, however, a single saturating flash applied at a fixed time (e.g., 10 min) after the actinic illumination can provide a useful approximation of the relative proportion of fast-relaxing photoprotective and slower-relaxing quenching, predominantly but not exclusively photoinhibition.

The slow-relaxing component NPQs is defined as

$$\mathrm{NPQs} = \frac{F_{\mathrm{m}} - F_{\mathrm{m\,rec}}}{F_{\mathrm{m\,rec}}} \tag{9}$$

where $F_{\mathrm{m\,rec}}$ is the recovered F_{m}, post-illumination. Fast-relaxing NPQf is then simply

$$\mathrm{NPQf} = \mathrm{NPQ} - \mathrm{NPQs} \tag{10}$$

An alternative approach has been developed by Ruban and Murchie [55] that aims to assay the extent of photodamaged PSII reaction centers. The basis of this is that, while photoinhibition results in quenching of F'_{m}, it *increases* F'_{o}. This contrasts with qE and other photoprotective processes, which quench both F_{m} and F_{o}. Previously, Oxborough and Baker [56] introduced a method to calculate F'_{o} in situations where it is not possible to measure this, particularly in fluorescence imaging systems. Their estimate of $F'_{\mathrm{o\,calc}}$ relies on the assumption that differences between F_{o} and F'_{o} are explained by qE. NPQ decreases F_{m} to F'_{m} to a larger extent than F_{o} to F'_{o}. However, the decrease in F_{o} becomes more significant while larger is the NPQ [56]. To consider this variability introduced by qE, the following equation was proposed in the calculation of F'_{o}:

$$F'_{\mathrm{o\,calc}} = \frac{F_{\mathrm{o}}}{\frac{F_{\mathrm{v}}}{F_{\mathrm{m}}} - \frac{F_{\mathrm{o}}}{F'_{\mathrm{m}}}} \tag{11}$$

Ruban and Murchie [55] compared measured and calculated values for F'_{o} and noted that the latter is often higher than the former. They argued that this results from photoinhibited PSII reaction centers. Based on this, they developed a parameter, $\mathrm{qP_d}$, to estimate the proportion of PSII reaction centers that return to the open state following illumination:

$$\mathrm{qP_d} = \frac{\left(F'_{\mathrm{m}} - F'_{\mathrm{o}}\right)}{\left(F'_{\mathrm{m}} - F'_{\mathrm{o\,calc}}\right)} \tag{12}$$

In the absence of photoinhibited reaction centers, the measured F'_{o} is the same as $F'_{\mathrm{o\,calc}}$ and $\mathrm{qP_d} = 1$. With increasing photoinhibition, $\mathrm{qP_d}$ falls, reflecting the loss of PSII reaction centers via photoinhibition.

5 To Modulate or Not to Modulate: Using Chlorophyll Fluorescence to Explore Plant Responses to Fluctuating Environments

Chlorophyll fluorescence has become one of the most powerful and used techniques to study photosynthesis. Although the physical phenomenon measured is the same (the yield of fluorescence), the use of different types of fluorometers has advantages and limitations that can make them more suitable for specific experimental approaches. Non-modulated fluorometers have been used extensively for phenotyping plants in the field [32, 57, 58]. The availability of cheap, user-friendly devices that measure the maximum quantum yield of PSII (F_v/F_m) allows for simple and fast estimations of the photosynthetic status of plants under control and stress conditions. This methodology is only limited by the capacity for dark incubation of the leaves prior to the saturating flash, which is easily overcome by using clips to cover the plant material. Several OJIP parameters have also been suggested for screening abiotic stress-tolerant varieties [30, 33]. However, although useful for characterizing phenotypic differences, the physiological meaning of these parameters needs to be considered carefully, due to the lack of a consensus on their interpretation [7, 23].

Modulated fluorometers (i.e., PAM fluorometers) are widely used for physiological studies aiming to get deeper insights into photosynthetic regulation. They allow for "real-time" estimations of photosynthetic performance, making it possible to follow energy partitioning at the level of PSII (non-photochemical dissipation and ΦPSII) and electron transport rates under different experimental conditions. Therefore, PAM fluorometers have been used to study regulatory processes and acclimation mechanisms to fluctuating environments, such as state transitions [59], NPQ induction and relaxation kinetics [54, 60], and high-light acclimation [61]. Although initially they were mainly associated with lab-based experiments, the development of portable devices has facilitated their use in the field [62, 63] or even for screening purposes [64]. The use of chlorophyll fluorescence in phenomics, remote sensing for precision agriculture, or net gross productivity of entire ecosystems is also an exciting field of research. This topic will be covered by another chapter in this book; for further discussion, refer [65].

6 Conclusion

Due to the central importance of photosynthesis for plant growth, studying its regulation is essential for understanding plant acclimation to a fluctuating environment. Within this context, the development in the 1980s of easy-to-use, now often portable, devices for the measurement of chlorophyll fluorescence has made this a fundamental method to assess photosynthesis. In this chapter, we have discussed the theoretical framework underlying this technique, focusing on non-modulated and modulated chlorophyll fluorescence analyses used in physiological experiments. We aimed to give a general overview of their main approaches implemented. Although the accessibility of these techniques has made them almost ubiquitous in plant physiology, there remains a danger that their complexities can be overlooked. We have presented the most common and widely used parameters describing photosynthetic energy partitioning at the level of PSII as well as other alternative parameters described in the literature. We hope that this chapter will help newcomers to the field address their questions in plant physiology using this powerful technique.

References

1. Mirkovic T, Ostroumov EE, Anna JM, Van Grondelle R, Govindjee, Scholes GD (2017) Light absorption and energy transfer in the antenna complexes of photosynthetic organisms. Chem Rev 117:249
2. Govindjee, Kern JF, Messinger J, Whitmarsh J (2010) Photosystem II. In: Encyclopedia of life sciences (ELS). John Wiley & Sons, Ltd., Chichester
3. Krause GH, Weis E (1991) Chlorophyll fluorescence and photosynthesis: the basics. Annu Rev Plant Biol 42(1):313–349
4. Lazár D (2015) Parameters of photosynthetic energy partitioning. J Plant Physiol 175:131–147
5. Maxwell K, Johnson GN (2000) Chlorophyll fluorescence – a practical guide. J Exp Bot 51(345):659–668
6. Kautsky H, Hirsch A (1931) Neue versuche zur kohlensäureassimilation. Naturwissenschaften 19(48):964
7. Stirbet A, Govindjee (2012) Chlorophyll a fluorescence induction: a personal perspective of the thermal phase, the J–I–P rise. Photosynth Res 113:15–61
8. Stirbet A, Govindjee (2011) On the relation between the Kautsky effect (chlorophyll a fluorescence induction) and photosystem II: basics and applications of the OJIP fluorescence transient. J Photochem Photobiol B 104(1–2):236–257
9. Pfündel E (1998) Estimating the contribution of photosystem I to total leaf chlorophyll fluorescence. Photosynth Res 56(2):185–195
10. Ruban AV, Wilson S (2021) The mechanism of non-photochemical quenching in plants: localization and driving forces. Plant Cell Physiol 62(7):1063–1072
11. Minagawa J (2011) State transitions – the molecular remodeling of photosynthetic supercomplexes that controls energy flow in the chloroplast. Biochim Biophys Acta Bioenerg 1807(8):897–905
12. Strasser RJ, Tsimilli-Michael M, Srivastava A (2004) Analysis of the chlorophyll a fluorescence transient. In: Chlorophyll a fluorescence. Springer, pp 321–362
13. Lichtenthaler HK, Rinderle U (1988) The role of chlorophyll fluorescence in the detection of stress conditions in plants. CRC Crit Rev Anal Chem 19(Suppl 1):S29–S85
14. Krause GH, Vernotte C, Briantais J-M (1982) Photoinduced quenching of chlorophyll fluorescence in intact chloroplasts and algae. Resolution into two components. Biochim Biophys Acta Bioenerg 679(1):116–124
15. Bradbury M, Baker NR (1981) Analysis of the slow phases of the in vivo chlorophyll fluorescence induction curve. Changes in the redox state of photosystem II electron acceptors and fluorescence emission from photosystems I and II. Biochim Biophys Acta Bioenerg 635(3): 542–551
16. Quick WP, Horton P (1984) Studies on the induction of chlorophyll fluorescence in barley protoplasts. I. Factors affecting the observation of oscillations in the yield of chlorophyll fluorescence and the rate of oxygen evolution. Proc R Soc Lond B Biol Sci 220(1220):361–370

17. Strasser RJ (1992) The Fo and the OJIP fluorescence rise in higher plants and algae. In: Regulation of chloroplast biogenesis. Springer, pp 423–426
18. Strasser RJ, Srivastava A (1995) Polyphasic chlorophyll alpha fluorescence transient in plants and cyanobacteria. Photochem Photobiol 61:32
19. Strasser RJ, Stirbet AD (2001) Estimation of the energetic connectivity of PS II centres in plants using the fluorescence rise O–J–I–P: fitting of experimental data to three different PS II models. Math Comput Simul 56(4):451–462
20. Delosme R (1967) Etude de l'induction de fluorescence des algues vertes et des chloroplastes au début d'une illumination intense. Biochim Biophys Acta Bioenerg 143(1):108–128
21. Stirbet A, Strasser RJ (1995) Numerical simulation of the fluorescence induction in plants. Arch Sci Genève 48:41–60
22. Schansker G, Tóth SZ, Strasser RJ (2005) Methylviologen and dibromothymoquinone treatments of pea leaves reveal the role of photosystem I in the Chl a fluorescence rise OJIP. Biochim Biophys Acta Bioenerg 1706(3):250–261
23. Garab G, Magyar M, Sipka G, Lambrev PH (2023) New foundations for the physical mechanism of variable chlorophyll a fluorescence. Quantum efficiency versus the light-adapted state of photosystem II. J Exp Bot 74(18):5458–5471
24. Samson G, Bruce D (1996) Origins of the low yield of chlorophyll a fluorescence induced by single turnover flash in spinach thylakoids. Biochim Biophys Acta Bioenerg 1276(2):147–153
25. Schreiber U (1986) Detection of rapid induction kinetics with a new type of high-frequency modulated chlorophyll fluorometer. In: Current topics in photosynthesis: dedicated to Professor L.N.M. Duysens on the occasion of his retirement. Springer, pp 259–270
26. Sipka G et al (2021) Light-adapted charge-separated state of photosystem II: structural and functional dynamics of the closed reaction center. Plant Cell 33(4):1286–1302
27. Oja V, Laisk A (2020) Time-and reduction-dependent rise of photosystem II fluorescence during microseconds-long inductions in leaves. Photosynth Res 145(3):209–225
28. Zushi K, Kajiwara S, Matsuzoe N (2012) Chlorophyll a fluorescence OJIP transient as a tool to characterize and evaluate response to heat and chilling stress in tomato leaf and fruit. Sci Hort (Amsterdam). 148:39–46
29. Gomes MTG, da Luz AC, dos Santos MR, do Carmo Pimentel Batitucci M, Silva DM, Falqueto AR (2012) Drought tolerance of passion fruit plants assessed by the OJIP chlorophyll a fluorescence transient. Sci Hort (Amsterdam) 142:49–56
30. Jedmowski C, Ashoub A, Brüggemann W (2013) Reactions of Egyptian landraces of Hordeum vulgare and Sorghum bicolor to drought stress, evaluated by the OJIP fluorescence transient analysis. Acta Physiol Plant 35:345–354
31. Rapacz M, Sasal M, Kalaji HM, Kościelniak J (2015) Is the OJIP test a reliable indicator of winter hardiness and freezing tolerance of common wheat and triticale under variable winter environments? PLoS One 10(7):e0134820
32. Bussotti F, Gerosa G, Digrado A, Pollastrini M (2020) Selection of chlorophyll fluorescence parameters as indicators of photosynthetic efficiency in large scale plant ecological studies. Ecol Indic 108:105686
33. Dos Santos VAHF, Nelson BW, Rodrigues JVFC, Garcia MN, Ceron JVB, Ferreira MJ (2019) Fluorescence parameters among leaf photosynthesis-related traits are the best proxies for CO2 assimilation in Central Amazon trees. Braz J Bot 42:239–247
34. Schreiber U, Schliwa U, Bilger W (1986) Continuous recording of photochemical and non-photochemical chlorophyll fluorescence quenching with a new type of modulation fluorometer. Photosynth Res 10(1):51–62
35. Roháček K, Barták M (1999) Technique of the modulated chlorophyll fluorescence: basic concepts, useful parameters, and some applications. Photosynthetica 37:339–363
36. Genty B, Briantais J-M, Baker NR (1989) The relationship between the quantum yield of photosynthetic electron transport and quenching of chlorophyll fluorescence. Biochim Biophys Acta Gen Subj 990(1):87–92

37. Kramer DM, Johnson G, Kiirats O, Edwards GE (2004) New fluorescence parameters for the determination of QA redox state and excitation energy fluxes. Photosynth Res 79(2):209–218
38. Björkman O, Demmig B (1987) Photon yield of O2 evolution and chlorophyll fluorescence characteristics at 77 K among vascular plants of diverse origins. Planta 170:489–504
39. Solhaug KA, Asplund J, Gauslaa Y (2021) Apparent electron transport rate – a non-invasive proxy of photosynthetic CO2 uptake in lichens. Planta 253:1–9
40. Clark JB, Lister GR (1975) Photosynthetic action spectra of trees: II. The relationship of cuticle structure to the visible and ultraviolet spectral properties of needles from four coniferous species. Plant Physiol 55(2):407–413
41. Holmes MG, Keiller DR (2002) Effects of pubescence and waxes on the reflectance of leaves in the ultraviolet and photosynthetic wavebands: a comparison of a range of species. Plant Cell Environ 25(1):85–93
42. Schöttler MA, Tóth SZ (2014) Photosynthetic complex stoichiometry dynamics in higher plants: environmental acclimation and photosynthetic flux control. Front Plant Sci 5:188
43. Bailey S, Horton P, Walters RG (2004) Acclimation of Arabidopsis thaliana to the light environment: the relationship between photosynthetic function and chloroplast composition. Planta 218:793–802
44. Chow WS, Melis A, Anderson JM (1990) Adjustments of photosystem stoichiometry in chloroplasts improve the quantum efficiency of photosynthesis. Proc Natl Acad Sci 87(19): 7502–7506
45. Johnson GN, Young AJ, Scholes JD, Horton P (1993) The dissipation of excess excitation energy in British plant species. Plant Cell Environ 16(6):673–679
46. Fryer MJ, Andrews JR, Oxborough K, Blowers DA, Baker NR (1998) Relationship between CO2 assimilation, photosynthetic electron transport, and active O2 metabolism in leaves of maize in the field during periods of low temperature. Plant Physiol 116(2):571–580
47. Edwards GE, Baker NR (1993) Can CO2 assimilation in maize leaves be predicted accurately from chlorophyll fluorescence analysis? Photosynth Res 37(2):89–102
48. Bilger W, Björkman O (1990) Role of the xanthophyll cycle in photoprotection elucidated by measurements of light-induced absorbance changes, fluorescence and photosynthesis in leaves of Hedera canariensis. Photosynth Res 25:173–185
49. Horton P, Ruban AV, Walters RG (1994) Regulation of light harvesting in green plants (indication by nonphotochemical quenching of chlorophyll fluorescence). Plant Physiol 106(2):415
50. Ruban AV, Johnson MP, Duffy CDP (2012) The photoprotective molecular switch in the photosystem II antenna. Biochim Biophys Acta Bioenerg 1817(1):167–181
51. Demmig-Adams B, Koh S-C, Cohu CM, Muller O, Stewart JJ, Adams WW (2014) Non-photochemical fluorescence quenching in contrasting plant species and environments. In: Non-photochemical quenching and energy dissipation in plants, algae and cyanobacteria. Springer, pp 531–552
52. Klughammer C, Schreiber U (2008) Complementary PS II quantum yields calculated from simple fluorescence parameters measured by PAM fluorometry and the saturation pulse method. PAM Appl Notes 1(2):201–247
53. Walters RG, Horton P (1991) Resolution of components of non-photochemical chlorophyll fluorescence quenching in barley leaves. Photosynth Res 27:121–133
54. Johnson GN, Young AJ, Horton P (1994) Activation of non-photochemical quenching in thylakoids and leaves. Planta 194:550–556
55. Ruban AV, Murchie EH (2012) Assessing the photoprotective effectiveness of non-photochemical chlorophyll fluorescence quenching: a new approach. Biochim Biophys Acta Bioenerg 1817(7):977–982
56. Oxborough K, Baker NR (1997) Resolving chlorophyll a fluorescence images of photosynthetic efficiency into photochemical and non-photochemical components–calculation of qP and Fv-/ Fm-; without measuring Fo. Photosynth Res 54:135–142

57. Poudyal D, Rosenqvist E, Ottosen C-O (2018) Phenotyping from lab to field–tomato lines screened for heat stress using Fv/Fm maintain high fruit yield during thermal stress in the field. Funct Plant Biol 46(1):44–55
58. Quamruzzaman M, Manik SMN, Livermore M, Johnson P, Zhou M, Shabala S (2022) Multidimensional screening and evaluation of morpho-physiological indices for salinity stress tolerance in wheat. J Agron Crop Sci 208(4):454–471
59. Taylor CR, van Ieperen W, Harbinson J (2019) Demonstration of a relationship between state transitions and photosynthetic efficiency in a higher plant. Biochem J 476(21):3295–3312
60. Kromdijk J et al (2016) Improving photosynthesis and crop productivity by accelerating recovery from photoprotection. Science 354(6314):857–861
61. Calzadilla PI, Song J, Gallois P, Johnson GN (2024) Proximity to photosystem II is necessary for activation of plastid terminal oxidase (PTOX) for photoprotection. Nat Commun 15(1):287
62. Edwards MS, Kim KY (2010) Diurnal variation in relative photosynthetic performance in giant kelp Macrocystis pyrifera (Phaeophyceae, Laminariales) at different depths as estimated using PAM fluorometry. Aquat Bot 92(2):119–128
63. Oivukkamäki J et al (2024) Field integration of shoot gas-exchange and leaf chlorophyll fluorescence measurements to study the long-term regulation of photosynthesis in situ. Tree Physiol 45:tpae162
64. Linn AI, Zeller AK, Pfündel EE, Gerhards R (2021) Features and applications of a field imaging chlorophyll fluorometer to measure stress in agricultural plants. Precis Agric 22(3):947–963
65. Sun Y et al (2023) From remotely sensed solar-induced chlorophyll fluorescence to ecosystem structure, function, and service: part I – harnessing theory. Glob Chang Biol 29(11):2926–2952

Non-photochemical Quenching in Plants: A Chlorophyll *a* Fluorescence Perspective

Rodrigo L. Gómez

Contents

Abstract Oxygenic photosynthesis, arguably the most important biological process on our planet, empowers phototrophic organisms to convert sunlight into the chemical energy that sustains ecosystems. Yet, this very process faces a constant challenge: capturing solar energy efficiently while avoiding the detrimental effects of

R. L. Gómez (✉)
Instituto de Investigaciones en Ciencias Agrarias de Rosario (IICAR), CONICET - Universidad Nacional de Rosario, Zavalla, Argentina

Cátedra de Fisiología Vegetal, Facultad de Ciencias Agrarias, Universidad Nacional de Rosario, Zavalla, Argentina

M. G. Lagorio (ed.), Comprehensive Series in Photochemical and Photobiological Sciences, Fluorescence of Living Plants 1, https://doi.org/10.1007/17361_2025_10
Published online: 5 July 2025

excess light. When light absorption outpaces the rate of carbon fixation, the inevitable consequence is a surge in reactive oxygen species (ROS), leading to photoinhibition and cellular damage. To combat this, plants and algae have evolved a sophisticated arsenal of photoprotective mechanisms, with non-photochemical quenching (NPQ) emerging as a central and highly regulated strategy. Chlorophyll *a* fluorescence has proven instrumental in deciphering these mechanisms, providing the foundation for powerful tools that unravel NPQ regulation and dynamics under changing environmental conditions. This chapter explores the fundamental principles underpinning NPQ and its pivotal role in enabling plants to thrive in dynamic light environments. Primary molecular mechanisms underlying NPQ and their key effectors are presented, alongside their physiological and ecological relevance. Existing controversies and knowledge gaps are also addressed, underscoring the need for continued research in this critical area. Furthermore, chlorophyll *a* fluorescence techniques, particularly pulse-amplitude-modulated (PAM) fluorometry, are highlighted as essential methods for investigating NPQ regulation and dynamics in response to environmental changes. Integrating these tools with molecular biology, microscopy, spectroscopy, and omics approaches provides a comprehensive understanding of how NPQ enables photosynthetic organisms to flourish in fluctuating light conditions. Understanding these mechanisms is essential for predicting plant responses to future climate conditions and for shaping sustainable agricultural practices that reinforce ecosystem resilience.

Keywords Non-photochemical quenching, Photoinhibition, Photoprotection, PSII energy partitioning

1 Introduction

1.1 "All That Glitters Is Not Gold": The Inevitable Dilemma of Phototrophs

Photosynthesis is the cornerstone of life on Earth, enabling phototrophic organisms, from cyanobacteria and algae to mosses and higher plants, to convert solar energy into chemical energy stored within organic compounds. During this process, chlorophylls absorb sunlight, fueling the chemical reactions that fix atmospheric carbon dioxide into sugars and the myriad other biomolecules vital for growth and metabolism. While photosynthesis underpins the energetic foundation of ecosystems, sustaining herbivores and higher trophic levels alike, its efficiency is intrinsically limited by a host of environmental factors [1].

Under intense light conditions, such as midday sunlight on clear days, the rate of energy absorption can outstrip the processing capacity of the photosynthetic apparatus, leading to its saturation. This surplus energy can be transferred to molecular oxygen, giving rise to reactive oxygen species (ROS) such as superoxide radicals

(O_2^-), hydrogen peroxide (H_2O_2), and singlet oxygen ($^1O_2$). ROS are highly reactive molecules capable of damaging proteins, lipids, and nucleic acids within chloroplasts, ultimately diminishing photosynthetic efficiency through light-induced stress, a phenomenon known as photoinhibition [2].

Photosynthetic organisms contend with an exceptionally dynamic light environment, characterized by fluctuations driven by diurnal and seasonal cycles, cloud cover, shading effects of neighboring vegetation, and even wind-induced leaf movement. To thrive amidst these constantly shifting conditions while simultaneously optimizing photosynthetic performance, plants have developed intricate regulatory mechanisms. These mechanisms strike a balance between efficient energy capture and robust protection against the damaging effects of excessive light. These photoprotective strategies are remarkably diverse, varying across different taxa and environmental contexts, and are influenced by factors such as nutrient availability, temperature variations, and water stress [3–5].

In plants, one line of defense against excess light involves optical screening, achieved through the accumulation of pigments like anthocyanins or the deposition of cuticle waxes on leaf surfaces. By absorbing or reflecting a portion of the incident light, these mechanisms reduce the amount of light that reaches the chloroplasts. This, in turn, alleviates the excitation pressure on chlorophyll molecules, mitigating the formation of harmful ROS [5]. Furthermore, structural adaptations within the photosynthetic apparatus play a significant role in photoprotection. For instance, adjustments in leaf orientation or overall morphology can minimize direct exposure to intense sunlight, reducing the flux of high-energy photons striking sensitive areas of the leaf surface. Beyond these immediate responses, plants also exhibit longer-term acclimation processes known as photoacclimation. These adjustments encompass the upregulation of stress-response proteins or the downregulation of pigment biosynthesis in response to prevailing environmental conditions, enabling plants to fine-tune their photosynthetic machinery over time to better cope with the ambient light environment [6, 7].

Another essential facet of photoprotection is the synthesis and action of antioxidant molecules and enzymes [8]. Antioxidant molecules including carotenoids, tocopherols, and ascorbate play a crucial role in scavenging ROS generated during periods of high-light exposure. A pool of free carotenoids resides within the thylakoid membrane, capable of directly neutralizing ROS through chemical reactions. Furthermore, certain carotenoids, most notably the oxygenated xanthophylls, are bound to various components of the photosynthetic machinery, where they not only act as ROS scavengers but also modulate the triplet chlorophyll yield ($^3Chl^*$) and regulate its lifetime, serving an important role in photoprotection [9]. Likewise, enzymatic components such as superoxide dismutase (SOD), catalase, and ascorbate peroxidase work in concert with nonenzymatic antioxidants like ascorbate and glutathione to neutralize ROS before they can inflict cellular damage. This comprehensive antioxidant defense system is critical for maintaining cellular integrity and ensuring the continued efficiency of photosynthetic processes, even under stress conditions [8].

Among the wide photoprotective mechanisms employed by phototrophs, non-photochemical quenching (NPQ) stands out as a pivotal process, enabling these organisms to dissipate excess absorbed energy as heat, thereby minimizing the production of ROS [10, 11]. While NPQ mechanisms are present across all plants and algae, some molecular differences exist among species [12]. As explained in the next section, by definition, NPQ encompasses a suite of processes that reduce chlorophyll *a* fluorescence yield without contributing to photochemistry, effectively and safely dissipating excess energy as heat [13]. Although the precise mechanistic details underlying NPQ remain a topic of ongoing discussion, its critical role in photoprotection and overall photosynthetic performance is widely recognized. Nevertheless, excessive or prolonged activation of NPQ can reduce photosynthetic efficiency by limiting the amount of light energy available for photochemical reactions [14, 15]. This trade-off underscores the importance of tightly regulated NPQ dynamics, which are crucial for optimizing plant productivity under variable environmental conditions. Furthermore, the kinetics of NPQ induction and relaxation are critical for plant survival in dynamic light environments, such as sunflecks in forest canopies or intermittent cloud cover, where rapid transitions between light intensities occur [16].

1.2 The Backstage of NPQ: Chlorophyll Excitation and Photochemistry

To fully explore NPQ mechanisms and their implications, it is essential to consider several key aspects of photosynthesis. This process begins with the absorption of light by specialized pigment-protein complexes, known as photosystems, embedded within thylakoid membranes. The precise control of light energy utilization is achieved at the molecular level through proteins that interact with or modulate these pigment-protein complexes [11, 17, 18].

Chlorophylls serve as the primary pigments responsible for capturing light energy. Upon absorbing photons of specific energy, electrons within chlorophyll molecules are promoted from their ground state to an excited state. In chlorophylls, these excited states typically involve transitions between singlet (S) and triplet (T) states, which differ in their spin multiplicity. In the S state, the excited electron retains the same spin orientation as in the ground state, while in the T state, the spin orientation is opposite to that of the ground state. In photosynthetic pigments, such as chlorophylls, the ground state is S, and excitation transitions primarily promote them to S excited states, since direct S-to-T transitions are theoretically forbidden at first order [19].

Accordingly, the absorption spectrum of chlorophyll reveals two significant bands: one in the red region, termed Qy, corresponding to the transitions to the first excited state (S1), and another in the blue region, associated with transitions to the second excited state (S2). However, owing to the rapid decay of electrons from higher excited states to S1, during which excess energy is dissipated as heat, both transitions ultimately result in occupancy of the lowest-level excited state S1. Chlorophyll in the S1 state is then rapidly deactivated back to the ground state (S0) through several pathways, including photon emission (fluorescence) or alternative non-radiative decay processes such as internal conversion (IC) and intersystem crossing (ISC). IC refers to a vibrational process that facilitates energy dissipation as heat without altering spin multiplicity. Conversely, ISC represents a transition from S to T states that involves a change in spin for the excited electron. Although formally forbidden, this transition can occur under specific conditions that lengthen the lifetime of the S1 excited state. Notably, isolated chlorophylls can exhibit ISC rates as high as 60%, leading to the formation of chlorophyll in its triplet excited state (^{3}Chl*), which can efficiently transfer energy to molecular oxygen (present in a triplet ground state, $^3O_2$), subsequently generating ROS. However, the absorption and deactivation properties of chlorophylls, and other pigments, are profoundly influenced by their molecular environment, which is a critical aspect for photosynthesis given that chlorophylls are densely integrated within the protein matrix of photosystems [20].

Photosystem I (PSI) and photosystem II (PSII) are sophisticated supramolecular complexes mediating the capture, conversion, and transfer of excitation energy in oxygenic photosynthesis. Within these complexes, chlorophylls can transfer excitation energy between pigments (chlorophylls and/or carotenoids) and also act as efficient electronic energy acceptors and donors [18]. They originated from a common ancestor, thus sharing some similar features, but have evolved into unique structures, components, and energy conversion strategies [21, 22]. Both photosystems are organized with a central photochemical reaction center (RC), where charge separation takes place, surrounded by an antenna system specialized in light harvesting and excitation energy transfer (EET) to the RCs (Box 1). The antenna surrounding each RC can be categorized into two groups: *core or inner antenna* and *peripheral or outer antenna*. The *core antenna* presents a fixed stoichiometry and is highly conserved among the different oxygenic photosynthetic organisms. On the other hand, the *peripheral antenna* has evolved independently to adapt to variable light intensities and spectral qualities found in different niches. The highly efficient transfer of excitation energy from antenna complexes to reaction centers is crucial for minimizing energy loss [21].

Box 1: Photosystem Organization

The functional architecture of the photosystems is central to understanding how fluorescence measurements can be used to infer the state of the photosynthetic apparatus.

Photosystem II (PSII) is a large, multi-subunit protein–pigment complex embedded in the thylakoid membranes of plants and algae. It exists as dimers, each consisting of two identical but independently functioning complexes containing a reaction center (RC) and the *inner antenna* proteins, CP43 and CP47. Within the PSII RC, a special pair of chlorophyll *a* molecules, known as

(continued)

Box 1 (continued)

P680, is coordinated by the D1 and D2 subunits and serves as the primary electron donor. RCs also include a suite of redox-active cofactors, pheophytin and plastoquinones, which mediate electron transfer through the complex. On the luminal side, the oxygen-evolving complex (OEC) is associated with the RC and catalyzes the splitting of water into oxygen, protons, and electrons, marking the first step of the photosynthetic electron transport chain. Into PSII supercomplexes a variable number of membrane-anchored peripheral antenna complexes are incorporated. Typically, each dimeric supercomplex contains two to four copies of trimeric LHCII complexes, and these dimers frequently further associate into higher-order megacomplexes or crystalline arrays, with several distinct structural types having been characterized. In plants, the light harvesting peripheral antenna system can be separated into two layers. The first one is composed of the monomeric antenna subunits, CP24, CP26, and CP29. The second layer is composed of trimeric LHCII proteins formed by the combination of three polypeptides, Lhcb1-3. Interestingly, Lhcb3 and CP24 diversified in parallel after emergence of land plants suggesting structural interactions among them [23].

In PSI, the RC is centered on a special pair known as P700, which consists of a dimer of *Chl a* molecules that absorb light at slightly longer wavelengths than PSII RC. The core complex of PSI comprises multiple subunits that coordinate various cofactors, including chlorophylls, carotenoids, and iron-sulfur clusters. These iron-sulfur clusters play a crucial role in facilitating rapid electron transfer steps that ultimately reduce ferredoxin, a mobile electron carrier essential for NADPH formation. In addition to its core complex, PSI features a peripheral antenna system, known as LHCI, with a belt shape, composed of four subunits, Lhca1-4. This antenna system is highly efficient at capturing photons and transferring energy to the reaction center. The organization of the antenna complexes in PSI is optimized to ensure high efficiency even under low light intensities, thereby maximizing the effective utilization of absorbed energy for electron transport [24].

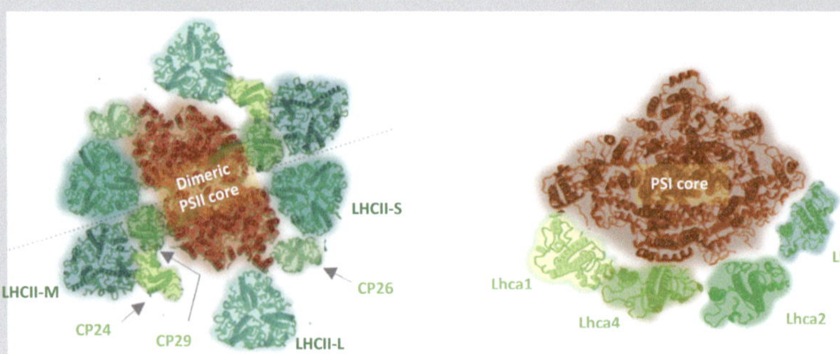

(continued)

Box 1 (continued)

The top view of both PSs is presented in the figure. Left, PSII dimer with a typical conformation found in low-light-acclimated plants with each PSII core dimer surrounded by its corresponding peripheral monomeric antennas and three LHCII trimeric complexes differing in their binding affinity indicated as S (strong), M (moderate), and L (loose) [21]. Right, PSI core surrounded by the belt-shaped Lhca monomeric antennas.

The linear electron transfer (LET) is initiated by the absorption of light energy by PSII, ultimately exciting P680 where charge recombination takes place, producing high-energy electrons. These electrons are then transferred to the plastoquinone (PQ) pool within the thylakoid membrane via the primary quinone acceptor Q_A and the secondary acceptor Q_B. Reduced plastoquinones (PQH_2) then diffuse through the thylakoid membrane to the cytochrome b_6f complex (Cyt b_6f), where it donates electrons while releasing protons into the lumen. Electrons are subsequently transferred to plastocyanin (PC), a mobile copper-containing protein that shuttles them to photosystem I (PSI). Within PSI, light energy absorbed by P700 leads to charge separation and transfer of electrons through a series of acceptors to reduce ferredoxin (Fd). Reduced Fd ultimately serves as reducing power for NADPH production by ferredoxin-$NADP^+$ reductase (FNR). This electron flow across the photosynthetic electron transport chain (PETC) is accompanied by the net translocation of protons from the stroma to the lumen of the thylakoids, establishing a proton gradient (ΔpH). This ΔpH, also referred to as the proton motive force (*pmf*), powers ATP synthase, an enzyme complex embedded in the thylakoid membrane, which synthesizes ATP from ADP and inorganic phosphate on the stroma side. Together, the ATP and NADPH generated in this photochemical phase provide the chemical energy and reducing power required for carbon fixation in the Calvin cycle [25].

2 Introducing the Cast: The NPQ Components

Photoprotection relies on NPQ as an evolutionary grounding that enables photosynthetic organisms, from aquatic algae to terrestrial plants, to dynamically cope with the excess light energy under fluctuating environmental conditions [26]. The term "non-photochemical quenching" derives from its defining feature: the quenching of chlorophyll fluorescence through mechanisms distinct from photochemical reactions in PSII. Specifically, NPQ arises from the non-radiative decay of singlet-excited chlorophyll states ($^1Chl^*$), which dissipates excess excitation energy as heat before it can drive harmful charge separations [27]. This process fulfills a critical role in preventing the overreduction of electron transport chains, thereby minimizing the formation of ROS and subsequent photooxidative damage.

A primary conceptual foundation of NPQ emerged from fluorescence kinetic studies using pulse-amplitude-modulated (PAM) fluorometry [28]. These studies revealed multiple NPQ components with distinct temporal and mechanistic signatures [29, 30]. Initially, a tripartite classification identified three main components: rapidly reversible energy-dependent quenching (qE), state transition-associated quenching (qT), and slowly reversible photoinhibitory quenching (qI). Further research has expanded this model, by redefining some of them and identifying additional components associated with pH-independent zeaxanthin quenching (qZ), chloroplast movement (qM), and sustained quenching (qH), each with unique induction and relaxation kinetics features [17]. These components operate through interconnected molecular pathways involving carotenoids, specific protein players, LHCII conformational changes, and macrostructure rearrangements of thylakoid membranes, though their relative contributions are not fixed.

NPQ manifests as a flexible integration of these subprocesses, modulated by species-specific adaptations and environmental variables such as light intensity, temperature, and hydration status among others [7, 15, 31]. This plasticity underscores the challenge of disentangling individual NPQ components in vivo, as overlapping signals and kinetic interdependencies can obscure their discrete contributions [17, 32–34]. Consequently, quantitative assessments require rigorous experimental designs that account for both temporal resolution (e.g., dark recovery intervals) and physiological context (e.g., abiotic stress history). The field continues to refine these methodologies, recognizing that the adaptive value of NPQ lies not in isolated mechanisms but in their synergistic capacity to balance photoprotection with photosynthetic efficiency under dynamic light regimes. Despite significant progress, discrepancies remain in our understanding of the exact molecular events that govern NPQ, and ongoing research continues to refine both the theoretical models and experimental methodologies in this field [6, 35]. This section delves into the main conceptual features of NPQ components and addresses some of the major discrepancies surrounding them.

2.1 Energy-Dependent Quenching (qE)

Energy-dependent quenching (qE) is the most prominent and dynamic component of NPQ, often being used interchangeably with "NPQ" due to its significant contribution to the overall process. The induction of qE typically occurs within seconds to minutes in response to illumination and consequent acidification of the lumen and can relax over several minutes once returned to low light or darkness [36, 37]. The term "energy-dependent" highlights its close relationship with the *pmf* resulting from *trans*-thylakoid ΔpH generated during electron transport [13]. Pioneer experimental approaches using ionophores have shown that qE is abolished without ΔpH, underscoring its essential role as a trigger for photoprotection. Besides this key factor, two other important players have been directly related to qE in land plants, a pH-sensitive protein (PsbS) and xanthophylls, mainly lutein (Lut) and zeaxanthin (Zea) [17, 36].

PsbS is a four-helix protein located within PSII belonging to LHC family, but without binding any pigment [38]. It contains two lumen-exposed glutamate residues that become protonated as the lumen pH decreases. This protonation induces a conformational change influencing the organization of pigment-protein complexes within the thylakoid membrane [16]. Intense research aimed at unraveling the molecular mechanisms behind PsbS was initiated from the discovery of PsbS-deficient mutant (*npq4*) in *Arabidopsis* [36]. Given that PsbS lacks pigment cofactors, it is unlikely to serve as the primary site for the quenching reactions that dissipate excited states of Chl [39]. However, the absence of PsbS does not impact non-photochemical quenching (NPQ) in other mutants lacking lutein and zeaxanthin accumulation, underscoring the essential role of PsbS in NPQ regulation [40]. Direct or indirect interactions between PsbS and LHCII have been proposed, primarily involving Lhcb1 subunits, while Lhcb2 and Lhcb3 are thought to play minor roles [32]. Additionally, the minor antenna subunit CP29 has been implicated in qE since its amplitude is affected in mutants lacking functional CP29, suggesting that its involvement may depend on PsbS-triggered reorganization of the thylakoid membrane [41, 42]. Although the role of PsbS in the quenching process, including its interaction with components like zeaxanthin and its potential function as either an essential element or a modulatory factor in qE, has been widely discussed, the precise molecular mechanisms involved remain elusive [33, 43].

As stated before, complete activation of qE also involves the xanthophylls, particularly Lut and Zea, which have been proposed to function both independently and synergistically to facilitate energy dissipation and/or as part of quenching sites [44]. In contrast, the other xanthophyll present in the antenna complexes, neoxanthin (Neo), seems to play a role as a sensor of the overall shift to a quenching state in antenna without directly contributing to energy dissipation [45].

The allocation of these pigments is predominantly into thylakoid membranes, where they are bound to LHC proteins. Typically, LHC monomers contain around 14 chlorophylls and coordinate 3–4 xanthophylls into specific binding sites. In LHCII, two of those binding sites, L1 and L2, coordinate Lut, while Neo is bound into the N1 site and an additional loose binding site in the periphery, V1, can bind violaxanthin (Vio) or Zea [46].

Lutein, the most abundant xanthophyll in land plants, is essential for structural stabilization of antenna proteins, light harvesting, and quenching of $^3Chl^*$ states. Experimental data using protein reconstitution and *Arabidopsis* mutants lacking lutein accumulation (*lut1* and *lut2*) suggest an essential role in stabilizing antenna proteins and trimerization of LHCII, most likely through its presence at the L1 binding site [47–49]. At both L1 and L2 sites, lutein can effectively transfer excitation energy to chlorophyll, thus participating in light harvesting [50]. Additionally, a critical role of Lut in photoprotection has been shown through its efficient quenching activity of proximal chlorophylls in triplet states at both L1 and L2 sites [44]. Moreover, Lut-dependent quenching is thought to be present also in monomeric LHCs [51].

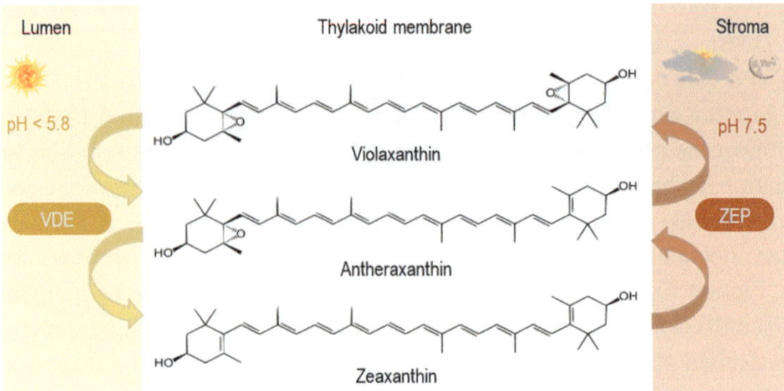

Fig. 1 The VAZ cycle. Under high light, acidification of the thylakoid lumen activates violaxanthin de-epoxidase (VDE), which sequentially converts Vio to antheraxanthin and then to Zea. Conversely, in low light or darkness, zeaxanthin epoxidase (ZEP) catalyzes the reverse reaction, regenerating Vio from Zea via the same intermediate

In contrast to Lut, which is typically present in a consistent quantity within photosystems, the content of Zea can vary significantly, and its accumulation correlates with the NPQ response in green plants [13, 52]. The xanthophyll cycle, or VAZ cycle, mediates the modulation of Zea in response to light intensity [44]. This pH-sensitive cycle dynamically regulates photoprotection through the light-dependent interconversion of Vio, antheraxanthin, and Zea (Fig. 1). This process optimizes energy dissipation while safeguarding photosynthetic efficiency under high-light conditions. However, Zea accumulation occurs more rapidly than its re-epoxidation back to Vio, reflecting a trade-off between photoprotection and photochemistry. For example, in *Arabidopsis*, full Zea-dependent qE induction is achieved within about 10 min under high light, whereas reversion to Vio in the dark or low light takes approximately 1 h.

Impairment in Zea accumulation due to lack of VDE activity in *npq1* mutant of *Arabidopsis* results in markedly reduced qE capacity. Conversely, *npq2* mutant, deficient in ZEP, exhibits constitutive Zea accumulation and a faster qE activation [13].

2.1.1 "To Quench or Not to Quench, That Is the Question": The Controversies Underlying qE Mechanism

The molecular architecture underlying qE remains a subject of active debate, particularly concerning the spatial distribution of quenching sites across the photosynthetic apparatus.

While early models localized qE primarily within LHCII aggregates [53], contemporary studies reveal a more complex landscape [33, 34, 54]. This multisite paradigm challenges the traditional view of LHCII-exclusive quenching, as biochemical analyses show that PsbS dynamically associates with both peripheral and core antennae depending on lumen pH [17, 32, 37, 51, 55]. Key findings from *Arabidopsis* mutants lacking LHCII demonstrate an important reduction in NPQ capacity, implicating these antenna complexes as major contributors to quenching [32, 56]. However, residual quenching persists even in LHCII-deficient systems, with structural and spectroscopic evidence pointing to additional sites in PSII where PsbS protein interactions facilitate protonation-dependent energy dissipation [42, 51]. Moreover, after protonation, a direct role of PsbS in quenching has been proposed by forming quenching sites with other antenna subunits and/or an indirect role by modulating conformational changes and reorganization of the thylakoid membrane macrostructure [32, 33, 37, 41, 42, 57, 58]. Despite the lack of consensus, in all proposed frameworks, the protonation of PsbS influences the lateral distribution of LHCII due to hydrophobic mismatch and electrostatic interactions at luminal sites, ultimately leading to heat dissipation [43].

The mechanistic indeterminacy extends to the photophysical nature of quenching. While xanthophyll-chlorophyll interactions are widely implicated, their exact role, as direct quenchers versus allosteric regulators of LHCII aggregation, remains unresolved [57, 59]. Several physical mechanisms underlying energy transfer between Chl and Car within PSII exist and can be grouped in three primary types of electronic interactions [60]. These mechanisms are not necessarily mutually exclusive and are proposed to collectively contribute to the efficient regulation of energy flow and photoprotection in photosynthetic complexes [17, 34].

One of these mechanisms involves a direct excitation energy transfer from Chl Q_y to the first excited singlet state of carotenoids (S1). This kind of process has been shown for Lut into L1 site of LHCII [61] and proposed, under the aggregation model framework, as the main pathway for energy dissipation generated within LHCII in the quenching state after conformational changes [62, 63].

The second important mechanism involves the formation of a carotenoid radical cation (Car^{+}), as a result of charge transfer (CT) between a Chl-Car pair, which quickly decays to ground state dissipating the energy as heat [64–66]. This CT mechanism received major attention since it can explain a direct role of Zea in energy quenching and has been proposed to be occurring mainly in minor antennas, particularly CP29, but absent in LHCII trimers [42, 51, 64, 67–69]. Studies on CP29 have identified a specific chlorophyll pair as critical for this process after replacement of Vio by Zea in monomeric Lhcb antenna complexes [42, 55, 70]. In support of the CT mechanism, transient zeaxanthin radical cations have been observed not only in the minor subunits involved in PsbS-dependent NPQ in higher plants but also in LHCSR-dependent quenching in algae, although similar signals are notably absent in the major trimeric LHCII complex [64, 66, 71].

In addition to the aforementioned mechanisms, excitation coupling between chlorophylls and carotenoids has been proposed promoting the formation of low-level dissipative electronic states that facilitate energy dissipation. In these processes, chlorophyll and carotenoids interchangeably act as quenchers or energy acceptors, depending on specific conditions and molecular configurations [69, 72, 73].

An early comprehensive model, known as aggregation model, was proposed over 30 years ago based on in vitro observations of fluorescence quenching in isolated LHCII aggregates induced by low detergent concentrations and/or low pH [74]. Thus, it was suggested that the aggregation of LHCII leads them to a dissipative state that quenches excess excitation energy. Central to this model is the concept that the conformational state of LHCII dictates its quenching capacity, a principle that has been maintained even as the model has evolved over time [63]. Consequently, under high light intensity, the aggregation model posits that LHCII undergoes conformational changes, leading to the formation of quenching clusters within thylakoid membrane. This process is enhanced by low pH and presence of zeaxanthin, conditions that mimic those found during qE activation [63]. Since aggregation has been proven not to be necessary for quenching, an inherent conformational change within the LHCII complex itself shifting it to a dissipative state has been proposed [75, 76]. Analysis of data from freeze-fracture electron microscopy and fluorescence recovery after photobleaching also suggest that an alteration in thylakoid membrane thickness observed during NPQ is a consequence of conformational changes in LHCII proteins, triggered by protonation and subsequent aggregation within the membrane [16, 77]. High-resolution cryo-electron microscopy and atomic force microscopy studies suggest that these changes alter the distances between pigments, influencing energy transfer dynamics and enhancing energy dissipation. Moreover, since the ability of chlorophyll-protein complexes to move within thylakoid membranes is essential for qE formation, factors that regulate this mobility are key determinants in the process, including specific minor light-harvesting proteins and PsbS. Also, membrane lipid compositions likely play pivotal roles shaping balance maintained throughout these dynamic transitions [78, 79]. Recent studies suggest that PsbS, influenced by the pH gradient, modulates LHCII clustering by interacting with the minor PSII antenna complexes and altering local membrane composition, modifying the interactions between LHCII and thylakoid lipids [80].

This current model clarifies that PsbS and Zea do not directly act as quenching agents but instead function as dynamic catalysts regulating NPQ through modulation of thylakoid membrane interactions. PsbS operates as an allosteric modulator, responding to lumen acidification by altering the structural organization of light-harvesting complexes (LHCII) within the membrane matrix. This protein facilitates reversible LHCII aggregation by lowering the energy barrier for conformational rearrangements, thereby enabling rapid transitions between light-harvesting and dissipative states. Concurrently, Zea, generated via the xanthophyll cycle, enhances membrane fluidity and stabilizes protein-protein interfaces in aggregated LHCII configurations. These synergistic effects create a metastable microenvironment that optimizes chlorophyll-chlorophyll or chlorophyll-carotenoid interactions for heat dissipation [63]. Critically, this regulatory framework decouples quenching site formation, governed by pigment-pigment dynamics, from its activation kinetics, controlled by PsbS/Zea-mediated macrostructural reorganization [81]. However, a significant limitation of the aggregation model was highlighted through experimental data, mainly from the analysis of *Arabidopsis* mutants. According to this model, the formation of aggregated LHCII-only domains should be the primary mechanism driving qE. Consequently, any mutations leading to LHCII overaccumulation or disruption of PSII supercomplexes would be expected to enhance NPQ activity. Instead, these mutations resulted in an increase in fluorescence, contradicting the predictions based on the aggregation model [17, 51].

In contrast with the previous model, solely depending on LHCII dynamics, alternative qE mechanisms have been proposed expanding quenching site(s) to other regions of PSII and also including a direct role of Zea and/or PsbS in energy-dependent NPQ [32, 51, 57, 68, 73, 82].

Recent advances in time-resolved fluorescence spectroscopy, complemented with conventional steady-state NPQ analyses, have led to the development of a comprehensive *four-state, two-site model* that reconciles some of the conflicting observations about qE dynamics in higher plants [44, 68, 69, 83]. This framework posits two distinct quenching sites, Q_1 and Q_2, operating through independent molecular mechanisms with distinct activation kinetics and regulatory dependencies. Q_1 is strictly PsbS dependent and is located within LHCII, while Q_2 is a PsbS-independent quenching site found within the minor antenna complexes attached to PS II. This Q_2 site is strongly dependent on zeaxanthin formation via the VAZ cycle and exhibits slower activation and relaxation kinetics. Since PsbS-dependent quenching is governed by the pH of the thylakoid lumen, it follows that the formation of Q_1 is controlled by lumen pH in a PsbS-dependent fashion. Moreover, experimental evidence suggests that Q_1 formation and relaxation closely parallel the buildup and dissipation of the thylakoid pH gradient over a 1–2-min timescale [83]. In contrast, the kinetics of Q_2 induction typically range from 10 to 20 min, consistent with the conversion of Vio to Zea, and its slower relaxation over more than 30–60 min, accompanied by Zea epoxidation. This observation suggests that Q_2 is involved in the more slowly inducible NPQ component, designated qZ [84], underscoring the complex interplay between quenching sites and the dynamic regulation of NPQ components.

Recent studies have further highlighted the importance of peripheral antennae in quenching, with CP29 and CP26 emerging as potential quenching sites due to their strategic positions within the PSII supercomplex [34, 42, 54, 70]. Additionally, bidirectional energy flow in the PSII supercomplex allows for efficient energy transfer between peripheral and core antennae, facilitating the activation of quenching centers under excess light conditions [73].

Current models thus converge on a hybrid mechanism: LHCII provides the primary quenching platform under transient light stress, while core-based processes become critical when peripheral antennae saturate or disassemble. This framework, however, lacks consensus on the hierarchy of site activation or their synergistic interplay during NPQ induction, a gap underscored by conflicting interpretations of mutant phenotypes and in vitro reconstitution studies [33, 51, 56, 85].

2.2 State Transition-Induced Fluorescence Quenching: qT

State transition-dependent quenching, qT, is a component of NPQ which operates on a timescale of 10–20 min. Unlike energy-dependent quenching (qE), which dissipates excess energy as heat, qT involves the redistribution of light harvesting PSII and PSI and thus is not a true quenching mechanism. However, it optimizes energy utilization under varying light conditions resulting in a decrease in fluorescence yield due to the stronger quenching capacity of PSI compared to PSII [86].

The origin of qT lies in the differential absorption spectra of PSII and PSI, which are influenced by their distinct pigment compositions. PSII antenna, primarily LHCII, is rich in *Chl b*, while PSI antenna contains more *Chl a* and some pigments are embedded in specialized scaffold contexts leading to the absorption in the far-red region [18]. This difference leads to PSII being preferentially excited by blue light and PSI by far-red light, creating a possible imbalance in excitation under non-saturating illumination. State transitions counteract this imbalance by redistributing a mobile pool of trimeric LHCII between PSII and PSI, thereby tuning the antenna cross sections of both photosystems [86–88]. This response is mediated by changes in the redox state of the plastoquinone (PQ) pool allowing optimizations of light absorption while minimizing the formation of reactive oxygen species under conditions of uneven excitation.

When PSII is overexcited, the reduced PQ (PQH_2) activates thylakoid-associated kinases (called STN7 in *Arabidopsis* and Stt7 in *Chlamydomonas*), which phosphorylate mobile LHCII, promoting their migration to PSI [89, 90]. These kinases phosphorylate specific N-terminal residues on LHCII proteins, reducing their affinity for PSII and promoting their migration towards PSI. Conversely, when PSI is preferentially excited, the oxidation of PQ leads to dephosphorylation of LHCII by the phosphatase TAP38/PPH1, allowing it to return to PSII [91, 92]. State 1 refers to mobile LHCII associated with PSII, whereas State 2 arises from the transfer of the additional LHCII to PSI, thereby augmenting its light-harvesting capacity. While the contribution of qT to total NPQ is generally low in land plants and negligible under saturating light conditions, it plays a more significant role in green algae like *Chlamydomonas reinhardtii* (Box 2).

Box 2: Divergent Strategies for Managing Excess Light in Algae, Mosses, and Higher Plants

All photosynthetic organisms share a wide array of strategies to mitigate photodamage, based on the mechanisms discussed in Sect. 2. However, the implementation of these strategies varies significantly across different taxa due to differences in inherent biological characteristics, habitat constraints, and evolutionary history [26].

While PsbS is a major player in land plant qE, the light-harvesting complex stress-related (LHCSR) protein is a critical component in the regulation of

(continued)

Box 2 (continued)

NPQ in mosses and eukaryotic algae, including diatoms and green algae. Structurally, LHCSR resembles other eukaryotic antenna proteins, featuring three transmembrane helices; however, in contrast to PsbS, it binds both chlorophylls and carotenoids [71]. Moreover, LHCSR is likely to possess flexibility in binding various carotenoids, although it does not bind neoxanthin. This adaptability implies that LHCSR can accommodate diverse carotenoid compositions, which can differ between species or change in response to environmental conditions [93].

The reliability for LHCSR-induced NPQ in algae is also sustained by the fact that the PsbS-encoding genes are absent in diatoms and red algae. Moreover, despite the presence of these genes in some green algae, the corresponding PsbS polypeptides have not been detected, indicating that these genes may be pseudogenes [94]. Mosses, like *Physcomitrella patens*, represent an evolutionary intermediate, utilizing both LHCSR- and PsbS-dependent NPQ. Knockout lines specifically depleted in LHCSR have shown that this protein is a primary component for reversible NPQ activity in mosses [95–97]. In contrast to algae, in *P. patens*, NPQ activity is also associated with PsbS, carrying the gene and accumulating the protein in thylakoid membranes. This suggests that PsbS might have been recruited during the transition of algae to land to enhance NPQ efficiency beyond what is achievable with LHCSR alone. Subsequently, PsbS-dependent mech anisms became the sole contributors to qE in vascular plants [94].

A key distinction between LHCSR and other antenna proteins lies in its intrinsic ability to dissipate energy as heat. This contrasts with PsbS, which, when inactive, has minimal influence on photon-harvesting efficiency, allowing higher plants to optimize light use in low-light conditions [98, 99]. Even in isolated form, LHCSR exhibits a short fluorescence lifetime, indicating efficient thermal dissipation of excitation energy [71]. The constitutive quenching activity of LHCSR is further enhanced under low luminal pH conditions, suggesting a pivotal role in thermal energy dissipation by receiving energy from nearby antenna complexes. This comprises a potential trade-off prompting these organisms to optimize their expression based on environmental conditions. Across various species, including the moss *P. patens*, the modulation of LHCSR accumulation in response to growth conditions appears to be a universal strategy for optimizing the amplitude of NPQ [100].

Beyond PsbS and LHCSR, carotenoids involved in xanthophyll cycles also exhibit specific features that vary among different taxa highlighting the diversity in NPQ regulation across species. In mosses, NPQ amplitude correlates with high levels of Zea in an LHCSR-dependent manner [40], where rapid energy transfer from excited chlorophyll to the carotenoid S1 state in LHCSR1 is operative [101]. Conversely, *C. reinhardtii npq1* mutant, which is unable to

(continued)

Box 2 (continued)

synthesize Zea, reaches similar levels of NPQ as wild type, suggesting that LHCSR quenching is independent of Zea but relies on Lut to serve a comparable function [71]. While in these free-floating algae, NPQ is Zea independent, biofilms forming algae exhibit a strong dependence on Zea [102]. Moreover, in diatoms, another cycle operates, similar to VAZ, mediating the conversion of diadinoxanthin to diatoxanthin by low luminal pH, which correlates with NPQ induction [103].

The evolution from unicellular algae to land plants also brought significant changes in thylakoid membrane organization including increased grana width and stacking. In land plants, grana consist of tightly stacked thylakoid discs, interconnected by non-appressed stroma lamellae. This highly structured network enhances the surface area for light capture and allows the spatial segregation of photosynthetic complexes, with PSII mainly located in grana and PSI located in margins and stroma lamellae. This arrangement contrasts with the simpler thylakoid systems in algae, where membranes are loosely appressed or arranged in smaller stacks. This organization, combined with the distinct localization patterns of LHCSR and PsbS, contributes to their functional differences. PsbS is permanently located in grana thylakoids, closely associated with PSII supercomplexes, whereas LHCSR is found in stroma membranes and grana margins [104]. These spatial arrangements influence their interactions with photosynthetic components and also their roles in other energy dissipation pathways as state transitions. In algae, state transitions are highly dynamic and closely integrated with LHCSR function, enabling rapid redistribution of excitation energy under transient light fluctuations. In contrast, the more rigid lateral organization of thylakoid membranes in land plants limits the amplitude of state transitions. Compared to a land plant such as *Arabidopsis* where only 15–20% of the PSII antenna is mobile, in *Chlamydomonas*, this number can reach up to 80% [86, 105]. This extensive mobility is crucial for algae, which often face hypoxic conditions that lead to overreduction of the PQ pool, necessitating robust state transitions for photoprotection [86]. Consequently, while the underlying qT mechanism is conserved, the quantitative impact on energy distribution and photoprotection is considerably more significant in microalgae than in higher plants. Instead, they rely on the rapid activation of qE by PsbS to adjust energy distribution, providing robust photoprotection against sustained high-light stress. Mosses occupy an intermediate position, retaining both systems, thus enabling them to dynamically balance excitation energy between PSII and PSI while engaging both inducible quenching by LHCSR and constitutive protection by PsbS [93].

Taken together, these differences reflect adaptations to varying light environments and highlight the evolutionary diversification of photoprotective strategies [12, 94].

2.3 Photoinhibition-Related Quenching: qI

The qI component of NPQ is a slow relaxing process, taking hours to days to recover, directly associated with the photoinhibition of PSII due to damage of the D1 protein at the reaction center [106, 107]. This leads to a decreased photochemical efficiency as a consequence of the reduction in the number of functional PSII reaction centers and the quenching in the inactivated centers, predominantly through charge recombination between Q_A^- and $P680^+$ [108]. The severity of photoinhibition, however, is determined by the active balance between the rate of photodamage and the efficiency of PSII repair mechanisms. Recovery from photoinhibition involves the degradation of the damaged protein and its replacement with newly synthesized D1. This PSII repair cycle includes reversible phosphorylation of PSII dimers, which prompts the monomerization of the damaged complexes and their translocation from the densely stacked grana regions to the more accessible stroma-exposed regions where repair occurs [109]. Under most conditions, the degradation of damaged D1 is fully compensated by the continuous PSII repair cycle. However, after prolonged high-light exposure, especially under environmental conditions that constrain photosynthetic performance, such as drought, low temperatures, or limitations in sink capacity, the qI component assumes a critical role. In these circumstances, photodamage tends to accumulate as the rate of damage exceeds the capacity for repair [110].

It is important to note that, whereas photoinhibition was traditionally viewed as a negative process, it has recently been proposed as part of an integrated protective strategy that preserves overall photosynthetic efficiency. Thus, photoinhibition not only serves to limit additional damage to PSII and reduce ROS generation, but also helps protect PSI, which lacks robust repair mechanisms [111, 112].

2.4 pH-Independent Zeaxanthin-Dependent Quenching: qZ

Zea-dependent quenching (qZ) is a crucial component of NPQ that operates on a timescale of 10–30 min and plays a crucial role in protecting the photosynthetic apparatus under prolonged high-light conditions. This slower induction and relaxation of quenching are tightly correlated with xanthophyll cycle functioning in response to high-light stress and Zea accumulation [84]. Unlike qE, which is dependent on a proton gradient and PsbS and is active in seconds to minutes after light imposition, qZ operates independently of low pH and can persist under dark conditions. Comparative analyses in *Arabidopsis* mutants have shown that qZ is independent of other NPQ components, marking it as a separate Zea-dependent mechanism. It was proposed that Zea binds mainly in the L2 site in minor antenna complexes and potentially peripheral sites in major LHCII trimers [84, 113]. The physiological significance of qZ is particularly evident under prolonged high-light exposure by functioning as a complementary mechanism to qE, ensuring photosynthesis optimization and protection not only under transient light fluctuations but also under prolonged, sustained light exposures. As such, qZ may act as a "light stress memory" within chloroplasts, enabling a rapid reactivation of photoprotective NPQ processes during recurrent periods of light stress [114].

2.5 Plastid Lipocalin (LCNP)-Mediated Sustained NPQ: qH

Recently, a sustained form of quenching, denominated qH, was described as a novel slowly relaxing component of NPQ. This type of quenching operates independently of several factors involved in other NPQ processes, including ΔpH, PsbS, and zeaxanthin. Instead, qH is mediated by specific molecular players LCNP, a lipocalin localized in the thylakoid lumen, and its modulators, suppressor of quenching 1 (SOQ1), a thylakoid membrane protein and RELAXATION OF QH1 (ROQH1), peripherally bound to the stroma lamellae membrane in the chloroplast stroma [115–117].

In response to adverse conditions, LCNP is thought to undergo oxidative modifications that trigger conformational changes in the antenna proteins. This transformation effectively switches light-harvesting complexes from a state of efficient energy transfer to one that favors heat dissipation. Under normal conditions, this protective mechanism is suppressed by SOQ1, but its inhibitory effect is alleviated under stressful conditions, possibly through redox-dependent alterations, thereby enabling photoprotection mediated by LCNP. Another key player conserved across the green lineage is ROQH1, which functions primarily as a qH relaxation factor. Through the use of mutants and overexpressors, it was demonstrated that qH fails to relax in *roqh1* mutants, whereas qH is absent in ROQH1 overexpressors. When mutations in *soq1* and *roqh1* are combined, qH becomes constitutive, as it cannot be either prevented or relaxed, leading to light-limited growth in *soq1 roqh1* plants. The antagonistic functions of LCNP and ROQH1 are dosage dependent, playing a critical role in protecting the photosynthetic apparatus and maintaining light-harvesting efficiency in plants [116].

Although both qH and qI contribute to slow relaxing NPQ, they operate through distinct mechanisms. Reflecting its alphabetical order, the label "H" for photoprotective quenching precedes "I." In contrast to qI, primarily associated with photoinhibition and involving charge recombination within inactive PSII centers, qH is an antenna-based mechanism dissipating the excess energy through thermal pathways [117]. A deeper insight into the interplay between qH and qI is essential for improving plant resilience under fluctuating environmental conditions. The knowledge could pave the way for strategies aimed at optimizing light harvesting while minimizing photodamage, with significant implications for crop productivity and bioenergy yields.

2.6 *Chloroplast Movement-Dependent Quenching: qM*

Recently, a new component of NPQ was proposed as being associated with chloroplast movement and the concomitant reduction of chlorophyll fluorescence yield, a phenomenon known as qM [118, 119]. Under illumination, a light-avoidance response decreases overall optical density by chloroplast migration within the cell to align parallel to the anticlinal cell walls [120]. This process is mediated by blue-light photoreceptor phototropin2, as first revealed through studies on *Arabidopsis* mutants *phot2*, which lack a functional receptor and exhibited a significant reduction in the fluorescence decay [121]. The repositioning leads to a "sieve effect" among chloroplasts, thereby lowering the number of excited fluorescence-emitting chloroplasts. This typically occurs over a period of 20–30 min and is not the result of an active quenching reaction per se but arises from reduced photon absorption. However, the relevance of qM in photoprotection to high light in plants grown under "shade conditions" has been questioned, suggesting that the role of chloroplast movements in total NPQ under these situations needs to be carefully evaluated [63].

3 Chlorophyll *a* Fluorescence: Illuminating Photoprotective Pathways Through Spectral Signatures

Since the early twentieth century, chlorophyll *a* fluorescence phenomenon has evolved into a field of photosynthetic research offering a window into the quantum ballet of energy partitioning within PSII by transforming transient signals into quantitative narratives of the fate of absorbed photons. Thus, these methodologies, as a glance through the interplay between photochemical quenching, thermal dissipation, and regulatory feedback loops, help to reveal how photosynthetic organisms balance light harvesting with photoprotective trade-offs. This approach has proven indispensable for dissecting NPQ mechanisms across scales, from isolated thylakoid membranes to intact leaves under field conditions [29, 30, 122–124]. Moreover, modern methodologies explore the spectral whispers of chlorophyll fluorescence to map photoprotective dynamics in vivo. Such precision has elevated fluorescence analysis beyond mere technique to an indispensable axiom for dissecting photoprotective strategies in oxygenic photoautotrophs.

3.1 *Pulse-Amplitude Modulation (PAM) Fluorometry Takes the Stage*

Traditionally, prompt chlorophyll fluorescence measurements relied on pre-darkening samples and data acquisition under carefully controlled light conditions, which inherently limited their applicability. While this approach remains invaluable for investigating the intricate photochemical processes within PSII, it is less suited for assessing PSII functionality under active illumination. Over time, advancements in fluorescence techniques have transformed this field, with pulse-amplitude modulation (PAM) fluorometry emerging as a groundbreaking innovation, a veritable Rosetta Stone, for deciphering the complexities of NPQ dynamics [28]. This method has enabled precise differentiation between photochemical and non-photochemical processes, offering a detailed analysis of PSII function. Furthermore, PAM fluorometry has significantly enhanced the resolution of NPQ components, providing profound insights into their kinetics, dynamic regulation, and underlying mechanisms [29, 30, 122, 123].

A fundamental principle underlying PAM fluorometry lies in its assumption that PSII RCs exist in two possible states, determined by their ability to achieve charge separation and deliver electron to the ETC. "Open" RCs are capable of initiating charge separation, whereas "closed" RCs are temporarily unable to perform this function. A defining feature of PAM fluorometry is the use of saturating light pulses (SP), which momentarily close all RCs by impairing photochemistry. This characteristic has earned the methodology its common association with SP analysis (Box 3). During an SP, photochemical energy conversion is deliberately suppressed, while non-photochemical processes remain active, providing alternative pathways for chlorophyll de-excitation.

Box 3: Measuring Photosynthetic Efficiency: PAM Methodology in the Nutshell

Given the pivotal role of PAM fluorometry in studying NPQ, understanding its technical features is essential. The technique is named after its primary characteristic: the use of a modulated light beam, known as measuring light (ML). This ML consists of very brief light flashes, typically lasting microseconds, applied at a specific frequency. A lock-in amplifier detector, modulated at the same frequency, measures the emitted fluorescence signal generated by these flashes. The intensity of ML is generally very weak, allowing it to probe fluorescence without significantly driving photosynthesis or perturbing the system. The methodological approach using PAM fluorometry to study NPQ is commonly known as *saturation pulse analysis*. This involves the use of an *actinic light* (AL), a continuous light source, usually red or white with user-set intensity, responsible for driving photosynthesis in the system under examination. Additionally, pulses of supra-saturating light, known as *saturation pulses* (SP), typically lasting less than a second, are used to transiently saturate photochemistry, temporarily eliminating photochemical quenching contributions.

Notably, only the fluorescence signal generated by ML is logged, while other light sources, such as AL and saturating pulses, are solely used to "perturb" the system under study. Thus, the significant advantage of using

(continued)

Box 3 (continued)
PAM fluorometry is its capacity to eliminate background light interferences, allowing measurements to be taken even under ambient light illumination.

This capability makes PAM fluorometry a versatile tool for assessing photosynthetic efficiency across various environmental conditions.

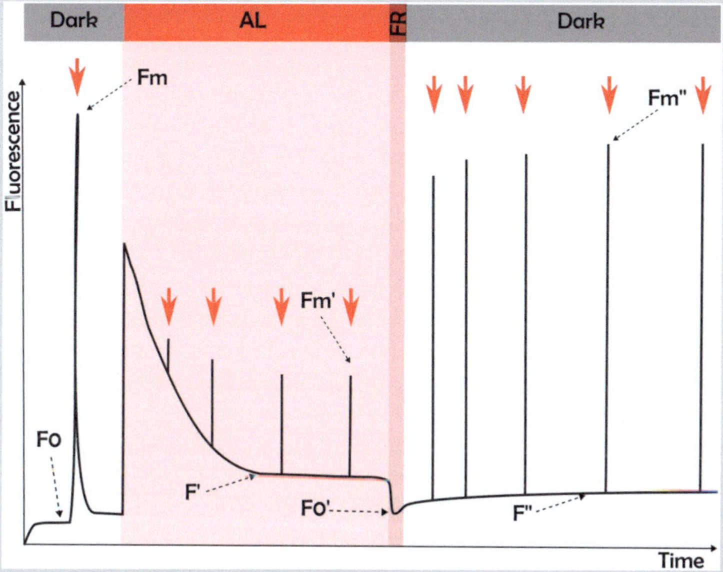

The figure illustrates a representative PAM trace obtained from quenching and relaxation analysis performed on a dark-adapted leaf. Measurements are initiated by turning on the ML and assessing two key parameters. The minimal fluorescence yield, F_o, is recorded prior to any illumination, representing the baseline fluorescence of the sample. Immediately afterward, an SP (indicated by red arrows in the figure) is applied to elicit the maximal fluorescence signal, F_m.

Subsequently, actinic light (AL) is activated, and additional SPs are applied at user-defined intervals. The fluorescence signals measured before each SP and during the SP are typically designated as F' and F'_m, respectively. Typically, the prime symbol (′) denotes fluorescence values obtained under AL illumination, while the double-prime symbol (″) is reserved for signals recorded during dark periods in relaxation analyses.

Modern commercial PAM fluorometers often incorporate a far-red (FR) light source to selectively excite PSI. During quenching analysis, brief FR illumination (1–5 s) is applied during transient AL interruptions to "drain"

(continued)

Box 3 (continued)
electrons from the ETC, effectively opening all reaction centers. This provides the minimal fluorescence signal under illumination, termed F_0'. However, under certain experimental conditions, determining F'_0 directly may prove challenging; thus, an alternative method for its estimation without using FR illumination has been developed [125].

Saturation pulse analysis is typically conducted in two primary modes: *quenching analysis*, performed under illuminated conditions, and *relaxation analysis*, carried out in darkness following a period of illumination [30, 124]. Both leverage chlorophyll fluorescence monitoring, but differ in their timing and objectives. *Quenching analysis* captures real-time insights into photochemical and non-photochemical processes, offering a snapshot of the photosynthetic system under light stress.

Conversely, *relaxation analysis* focuses on the recovery of the photosynthetic system during an extended dark period after illumination. While *relaxation analysis* offers a more comprehensive examination, it is significantly more time consuming, making it less practical for large-scale or high-throughput experiments. Therefore, the choice of methodological approach depends on specific research objectives. As detailed in Chap. 2 and supported by the cited literature, PAM fluorometry provides a wide range of measurable parameters. Nonetheless, this section will concentrate specifically on those parameters providing direct insights into NPQ mechanisms, elucidating their respective roles and interrelationships.

One of the earliest parameters developed for quantifying NPQ using PAM fluorometry is qN which was formulated in analogy to photochemical fluorescence quenching, qP [28, 107]. The equation is given by

$$\mathrm{qN} = \frac{\left(F_{\mathrm{v}} - F'_{\mathrm{v}}\right)}{F_{\mathrm{v}}} \tag{1}$$

where F_{v} and F'_{v} are the differences between maximal and minimal fluorescence measured under dark and light conditions, respectively. As a dimensionless value ranging from 0 to 1, qN offers a relative measure of NPQ. However, qN exhibits limited sensitivity to changes in non-photochemical processes when they are already highly engaged and of considerable magnitude. Moreover, accurately determining F'_{o}, a necessary component of the qN calculation, can be challenging in certain experimental settings [125].

Later, the most famous parameter, *NPQ*, was introduced based on comparing the maximum fluorescence yield in dark-adapted leaves (F_{m}) to the maximum fluorescence yield in leaves exposed to actinic light (F'_{m}) [126]. To differentiate the parameter from the general non-photochemical quenching processes in this chapter, it will be presented in italics:

$$NPQ = \frac{F_m - F'_m}{F'_m} \tag{2}$$

This approach draws from the Stern-Volmer principle, which establishes a linear relationship between the degree of fluorescence quenching and the concentration of the quenching agent in solution. Although the quenching environment of chlorophyll within the highly structured pigment-protein complex of a photosystem differs significantly from that of chlorophyll in solution, applying this principle to saturation pulse analysis of *Chl a* fluorescence has proven to be a valuable method for quantifying *NPQ* [29, 30, 124]. The strength of the *NPQ* parameter lies in the fact that any observed differences between F_m and F'_m can solely be attributed to non-photochemical processes, as photochemistry is transiently saturated during the application of saturating pulses. Moreover, *NPQ* remains one of the most widely used parameters derived from PAM fluorometry until now. Compared to the qN parameter, *NPQ* offers several advantages: it does not require simultaneous measurement of F'_o, it exhibits greater sensitivity, and it is not arithmetically constrained to values below 1.

Traditional NPQ measurements require determining maximum fluorescence (F_m) in fully dark-adapted samples, where NPQ is assumed negligible. However, this dark-adaptation requirement presents several limitations hampering high-throughput studies, field experiments, and imaging applications. Furthermore, the complexity of these measurements is compounded by potential artifacts arising from incomplete quenching relaxation, photodamaged centers, chloroplast movements, and inconsistencies when comparing values across different time points. To address these limitations, a new fluorescence parameter, NPQ_T, was developed to estimate *NPQ* without requiring dark adaptation, by reformulating the equation to use F'_o and assuming a set maximal quantum efficiency [127]:

$$NPQ_T = \left[\frac{4.88}{\frac{F'_m}{F'_o} - 1}\right] - 1 \tag{3}$$

NPQ_T can be applied during steady-state illumination, even in field conditions, and is suitable for imaging applications where samples move, although they require fluorometers with far-red sources to accurately calculate F'_o. Since this approach overcomes the limitations of the standard method, it became highly appealable for high-throughput exploration [128–130]. Moreover, modern devices designed for large-scale high-throughput plant phenotyping increasingly rely on this methodology [131].

Another modern methodological approach based on SP analysis has been developed focused on determining the photoprotective NPQ (pNPQ), which specifically safeguards PSII by minimizing the risk of photoinhibition [132]. This approach builds upon traditional quenching analysis techniques but offers a more nuanced understanding of the protective role of NPQ compared to traditional methods. Central to these developments is the parameter qP_d, which measures photochemical quenching in the dark immediately after light exposure. Thus, it provides a functional assessment of PSII integrity by evaluating the extent to which reaction centers remain open and operational [132–134]. Its calculation is based on the traditional qP parameter using the actual measured minimal fluorescence in the dark immediately after an illumination period ($F'_{\mathrm{o,act}}$) and its calculated theoretical value ($F'_{\mathrm{o,calc}}$) [125].

$$\mathrm{qP_d} = \frac{F'_{\mathrm{m}} - F'_{\mathrm{o,act}}}{F'_{\mathrm{m}} - F'_{\mathrm{o,calc}}} \tag{4}$$

Experimental validation of these methods has primarily involved various *Arabidopsis* genotypes that either lack or overexpress key components of non-photochemical quenching (NPQ), as well as chemical disruption of proton gradients across thylakoid membranes. Future studies will benefit from integrating these methodologies with broader metrics of photosynthetic performance, such as electron transport rates and carbon assimilation efficiency, thereby providing a more comprehensive understanding of how NPQ impacts overall photosynthetic function and plant productivity [133, 135–137].

In the investigation of NPQ mechanisms, *relaxation analysis* is frequently employed following an illumination period to assess the contributions of distinct NPQ components. A range of methodologies have been implemented, grounded on the direct analysis of temporal dynamics in F''_{m} levels and/or associated mathematical formalisms. This analytical approach facilitates the empirical discrimination of NPQ components, based on variances in their relaxation kinetics under dark conditions. However, it is pertinent to acknowledge that relaxation times are empirically derived and can vary across species and sample treatments as well as be affected by the duration and intensity of prior actinic illumination [20, 138]. Thus, the methodological approach should be carefully selected based on the system under investigation.

3.1.1 PSII Energy Partitioning and Its Relation to Non-photochemical Quenching

Since the mid-1990s, the study of energy partitioning in photosynthesis has advanced significantly, with early efforts focusing on the overall distribution of absorbed energy [139]. However, more recent methodologies have refined this framework by incorporating distinctions between different NPQ components [140–142]. The implementation of relaxation analysis further enriched this conceptual model by enabling a more comprehensive evaluation of energy partitioning parameters and their temporal dynamics [20, 138, 143].

The concept of energy partitioning in PSII is rooted in the principle of energy conservation, fundamentally grounded in the fate of absorbed energy, which is distributed among photochemistry, fluorescence emission, and thermal dissipation. Chlorophyll fluorescence analysis, particularly via PAM-based SP analysis, provides a direct quantification of the proportion of light energy allocated to each of these critical processes. In this context, quantum yield parameters, typically denoted by a capital Y or the Greek letter Φ, represent the fraction of absorbed light energy funneled into each specific process.

Within this conceptual framework, the fraction of energy utilized for photochemistry, assessed by ΦPSII, represents the proportion of light energy captured by PSII that is directed towards driving electron transport. Conversely, ΦNPQ quantifies the yield of regulated thermal dissipation through light-induced NPQ mechanisms. Meanwhile, ΦNO accounts for a minimal portion of energy channeled by non-regulated dissipative pathways, also known as constitutive non-photochemical dissipation. The sum of these quantities equals unity, ensuring a complete account of the absorbed light energy as required by the underlying concept of energy conservation (Fig. 2a).

Examining the relative contributions of each process offers valuable insights into the photoprotective mechanisms that plants employ to optimize their photosynthetic performance. Building on PAM fluorometry, various equations for calculating quantum yields have been proposed by different authors. A comprehensive review of these equations, along with the theoretical assumption underpinnings, has been provided by Lazár [138]. Despite minor differences, the equations are fundamentally consistent, allowing for practical interconversion. Therefore, the simplest forms of these equations are presented in Fig. 2a for clarity and ease of application.

Using energy partitioning, the adjustment and balance of photosynthetic processes can be followed under different light and stress conditions. For instance, under high-light stress, an increase in ΦNPQ is accompanied by a decrease in ΦPSII, highlighting the shift towards photoprotection at the expense of photochemistry. However, a reduction of ΦPSII without modification of ΦNPQ leads to higher values of ΦNO. This translates to an overall enhancement of energy dissipation

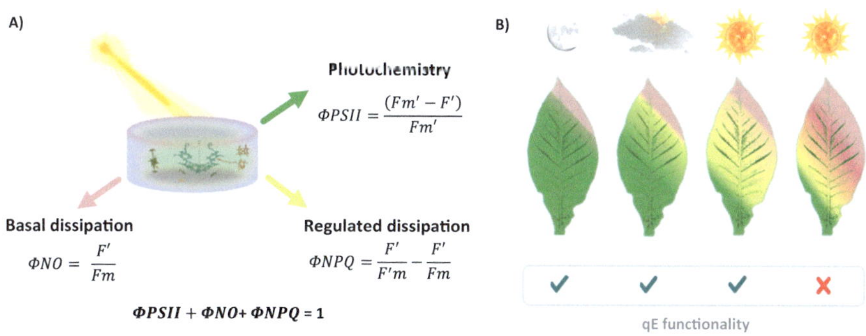

Fig. 2 The fate of energy into PSII. (**a**) Schematic depiction of energy partitioning in PSII after light absorption and equations for estimations of the corresponding quantum yields based on PAM fluorometry. (**b**) Graphical representation of PSII energy partitioning in leaves with completely active or impaired qE capacity (as indicated at the bottom) under different light conditions (as indicated on top). The colored areas in the leaves illustrate the proportion of quantum efficiency for photochemistry (green, ΦPSII), regulated NPQ processes (yellow, ΦNPQ), and nonregulated non-photochemical dissipation (red, ΦNO)

through nonregulated intrinsic mechanisms, including ISC, leading to higher probability of energy transfer to molecular oxygen and consequent generation of ROS (Fig. 2b).

The choice between energy partitioning parameters and the traditional parameters depends on the specific goals of the study. For detailed investigations requiring precise differentiation between photoprotective mechanisms and damaging processes or insights into PSII efficiency, energy partitioning parameters are more informative but demand advanced techniques and expertise. On the other hand, *NPQ* remains a reliable and accessible tool for general assessments of photoprotection through heat dissipation. Moreover, this parameter is mathematically related to quantum yields of regulated and nonregulated NPQ (Eq. 5), illustrating their interconnection and interdependence [138]:

$$NPQ = \frac{\Phi \mathrm{NPQ}}{\Phi \mathrm{NO}} \tag{5}$$

Incorporating relaxation kinetics assessments during a post-illumination dark period provides valuable insights into the complex mechanisms underlying NPQ. A pioneer approach sought to divide light-induced NPQ yield based on its origin, separating ΦNPQ into a fraction corresponding to energy-dependent quenching (ΦqE) and another reflecting contribution from state transitions and photoinhibition (ΦqT + qI). Applying this framework to *Arabidopsis npq1* mutants by using a fixed 10-min threshold for complete qE relaxation revealed a correlation between qE and time-resolved spectroscopy data, supporting a main role of charge-transfer quenching involving zeaxanthin cation radicals (Zea^+) in qE [143].

Another approach adopted a less stringent separation, categorizing light-induced NPQ components as either fast or slow based on relaxation analysis. The threshold time distinguishing these components varied by species, with 1 min for *Arabidopsis* and 5 min for rice. Quantum yields were then derived for the fast (ΦNPQ_{fast}) and slow (ΦNPQ_{slow}) NPQ components [144]. A more comprehensive partitioning of NPQ components was later proposed, based on dark relaxation kinetics without preselecting relaxation thresholds. By plotting maximal fluorescence during dark recovery against time, this method identified two distinct slope changes corresponding to 2 and 20 min, attributed to the relaxation of qE and qT, respectively. These observations led to a division of ΦNPQ into three distinct quantum yield components: ΦqE, ΦqT, and ΦqI [145].

In summary, using dark relaxation within an energy partitioning framework offers a more detailed resolution of NPQ components. However, the duration of the relaxation periods is somewhat arbitrary and should be carefully selected based on the specific research objectives. Furthermore, more complex energy partitioning theoretical frameworks have been developed, considering the role of photoinactivated reaction centers or sustained NPQ in evergreens [146, 147]. Since no single approach can be considered universally superior across all experimental scenarios, its selection should be optimized to capture desired facets of photosynthetic regulation [138].

3.1.2 Applications and Caveats of PAM Fluorometry in Assessing NPQ

As PAM-based chlorophyll fluorescence analysis tools become increasingly accessible, the need for precise protocols and thoughtful data interpretation has increased substantially. This heightened importance stems from the fact that, although generating fluorescence signals is relatively simple, ensuring their reliability and interpretability demands meticulous methodological attention to prevent common errors [20, 148].

A critical step in assessing NPQ processes is the determination of maximal fluorescence, both in dark-adapted (F_m) and illuminated (F'_m) conditions. However, this measurement poses several challenges. The F_m measurement can be influenced by the duration of dark adaptation and preexisting quenching resulting from environmental stresses. Given that certain NPQ processes exhibit prolonged relaxation times, their contributions should be considered when establishing appropriate dark-adaptation conditions. Moreover, the intensity and duration of saturation pulses are crucial, as they must ensure complete reduction of the quinone electron acceptor (Q_A) in PSII to obtain accurate PAM-related parameters [148, 149].

During illumination, F'_m determinations can be affected by the overly frequent application of saturating pulses. Furthermore, achieving complete reduction of Q_A can be challenging due to the rapid turnover of PSII acceptor pools, even when employing high-intensity saturation pulses [150]. Such high-intensity pulses, however, may trigger additional quenching reactions, including reaction center quenching, chlorophyll cation formation, and chlorophyll triplet-state formation. These auxiliary reactions can be influenced by the pH of the thylakoid lumen and may occur more frequently with increasing pulse intensity and duration [151, 152]. To mitigate these challenges, a multiple-phase flash (MPF) method has been proposed. The MPF, lasting approximately 1 s, involves a gradual stepwise decrease in irradiance. By analyzing the relationship between apparent F'_m and irradiance within each step of the MPF, it then becomes possible to estimate F'_m at effectively infinite irradiance. This approach is particularly valuable under high-actinic-light conditions, providing a more accurate and reliable assessment of photosynthetic performance [153].

Another assumption inherent in PAM-derived parameter calculations is that fluorescence signals primarily originate from PSII at room temperature. However, studies have shown that the contribution of PSI to the F_o signal can be substantial, reaching approximately 30% in C3 plants and up to 50% in C4 plants at wavelengths above 700 nm [154]. This issue is exacerbated by the fact that most commercial fluorometers, in an effort to avoid technical challenges, measure fluorescence by integrating intensity at wavelengths greater than 700 nm, inadvertently amplifying the relative contribution of PSI fluorescence. A specific method has been developed to address and correct for this contribution in F_o determinations when needed [155]. Furthermore, under certain conditions, the contribution of PSI fluorescence can be significantly altered, such as through direct energy transfer from PSII to PSI, or "spillover," proposed as a mechanism of NPQ in various organisms, including plants, algae, cyanobacteria, and lichens [156, 157].

Other biological processes such as state transitions, chloroplast movements, and absorptive effects from leaf surface features, such as trichomes and cuticular waxes or epidermal pigments, can also introduce artifacts in fluorescence measurements. These factors alter the effective absorption cross section of PSII antennae and/or interfere with signal detection, complicating the interpretation of fluorescence data [148].

Therefore, accurate NPQ measurement demands careful consideration of both biological factors and instrumental limitations. Addressing potential sources of error within the experimental setup is paramount to obtaining meaningful insights into photosynthetic energy dissipation mechanisms. Integrating NPQ data with complementary physiological measurements and computational modeling significantly enhances interpretive accuracy, leading to a more comprehensive understanding of photoprotection dynamics in plants [31, 122, 123, 148, 149, 158].

3.2 Beyond PAM Fluorometry: A Glimpse into Alternative Chl a *Fluorescence Techniques for Investigating NPQ*

While PAM fluorometry remains a cornerstone for studying non-photochemical quenching (NPQ) and photoprotection, several advancements in techniques based on *Chl a* fluorescence have expanded the toolbox available for exploring photosynthetic regulation. These alternative methods differ in their excitation sources, detection strategies, and measurable parameters, offering complementary insights into energy dissipation mechanisms and photosystem dynamics.

One widely used approach is **low-temperature (77 K) fluorescence spectroscopy**, which enables the separation of emissions from PSI and PSII based on their distinct spectral signatures. At room temperature, variable fluorescence is dominated by PSII, with minimal contribution from PSI. However, at 77 K, fluorescence from both photosystems becomes distinguishable due to their characteristic emission maxima. PSII exhibits peaks at 685 and 695 nm, while PSI emits around 730 nm, with slight variations depending on the species [159]. This spectral differentiation facilitates detailed analysis of the antenna systems associated with both photosystems. Thus, this technique is particularly valuable for studying state transitions, where changes in the relative amplitudes of PSII and PSI emission bands indicate shifts in energy distribution between the two photosystems. Although inherently static, low-temperature measurements can reconstruct quenching kinetics by freezing samples at different time points during light treatment or dark recovery, providing insights into dynamic processes.

Time-resolved fluorescence spectroscopy (TRFS) is another powerful technique that measures the decay of chlorophyll fluorescence following excitation with a short laser pulse. By resolving fluorescence lifetimes, TRFS provides detailed information about excited-state dynamics and interactions between pigments such as chlorophylls and carotenoids. This makes it particularly useful for dissecting how these interactions are modulated by NPQ mechanisms [160]. A more detailed discussion of TRFS can be found in Chap. 5 of this book.

Fast repetition rate fluorometry (FRRF) employs rapid sequences of saturating light pulses to resolve fast kinetics of photosynthetic electron transport and antenna connectivity at PSII. This method offers valuable insights into factors influencing NPQ [161]. A derivative technique known as **light-induced fluorescence transient (LIFT)** uses subsaturating excitation flashlets to monitor chlorophyll fluorescence induction and relaxation dynamics [162, 163]. In addition to providing information similar to PAM fluorometry, LIFT measures downstream electron transport rates beyond Q_A and offers the unique advantage of enabling remote measurements. This capability bridges the gap between leaf-level and canopy-scale analyses, making it particularly useful for ecological studies [164, 165].

Among emerging techniques, **solar-induced fluorescence (SIF)** has gained significant attention as a noninvasive method for assessing photosynthetic activity at larger scales. SIF utilizes natural sunlight as an excitation source and measures fluorescence emitted by vegetation remotely using hyperspectral sensors mounted on drones, aircraft, or satellites. This approach has proven invaluable for monitoring photosynthetic efficiency across ecosystems and linking canopy-level fluorescence to environmental stressors. SIF is projected to become one of the most widely used platforms for chlorophyll fluorescence studies due to its scalability and relevance for climate change research [166].

Finally, **fluorescence lifetime imaging microscopy (FLIM)** has emerged as a cutting-edge tool for visualizing spatial variations in NPQ at cellular or subcellular levels. By mapping fluorescence lifetimes across tissues or organelles, FLIM provides unprecedented detail about energy dissipation dynamics within individual chloroplasts or thylakoid membranes [167].

Each of these techniques offers unique strengths tailored to specific research questions. Combining data from multiple approaches enhances interpretive accuracy and provides a comprehensive understanding of photoprotection dynamics across scales – from molecular interactions within photosystems to ecosystem-level responses. Regardless of the method employed, careful consideration of its limitations and validation through complementary techniques remain essential for robust conclusions.

4 Coordinating Light Energy Dissipation with Metabolic Processes in Photosynthesis

4.1 Cellular Energy Budget Regulation and NPQ

NPQ is part of a sophisticated energy balancing network integrating light energy capture, its partitioning between productive photochemistry and dissipation via non-photochemical quenching (NPQ), and the matching of energy supply to fluctuating metabolic demands. Under conditions when the supply of ATP and NADPH exceeds the immediate needs of the cell, the system engages protective mechanisms. These are primarily triggered by the acidification of the thylakoid lumen, leading to the rapid induction of qE and the downregulation of electron flow at the cytochrome b_6f complex to prevent PSI photodamage [25, 168]. The modulation of lumen pH plays an essential role in the interplay between these processes, ensuring a dynamic balance that safeguards photosynthetic efficiency.

Globally, the efficiency of electron transport is tightly regulated to balance the production of ATP and NADPH with the metabolic demands of the cell. While LEF simultaneously generates ATP and NADPH, the stoichiometric requirements of the metabolic processes within chloroplasts are very variable, particularly under conditions of environmental stress or specific photosynthetic adaptations [169]. Cyclic electron flow (CEF) provides a mechanism to meet these demands by selectively driving *pmf* formation without producing NADPH [168]. This process involves electron recycling from Fd to the PQ pool, followed by oxidation at the cytochrome b_6f complex, enabling precise tuning of ATP/NADPH ratios. Under non-stress conditions in C3 plants, CEF operates minimally as LEF meets ATP demands but is upregulated during high ATP demand. Beyond balancing ATP/NADPH budget, CEF regulates photosynthesis by enhancing thylakoid lumen acidification. The ΔpH component of *pmf* generated by CEF slows plastoquinol oxidation at the Qo site of Cyt b_6f complex, a process known as "photosynthetic control," preventing overreduction of PSI and contributing to redox homeostasis. Also relevant to the acidification process is the regulation of the activity of the ATPase, which becomes reduced when levels of CO_2 drop, slowing proton efflux from the thylakoid lumen and leading to enhanced *pmf* [170]. Therefore, the induced lumen acidification activates NPQ by promoting VAZ cycle and PsbS protein protonation.

Complementing chloroplast-based adjustments, the malate valve operates as an essential mechanism for energy balancing by shuttling reducing equivalents from the chloroplast to the mitochondria. This process converts excess NADPH into mitochondrial ATP via malate oxidation, effectively trading reducing power for ATP [171]. The synergistic interplay among NPQ, CEF, and malate valve affords photosynthetic organisms with remarkable flexibility under varying light intensities and dynamic metabolic demands.

Under low-light conditions, the system prioritizes high-efficiency energy conversion and conversely, when the risk of photodamage escalates under excess light, a larger fraction of the absorbed energy is diverted to NPQ and CEF, the latter serving as a rapid, adjustable stopgap to bridge transient imbalances in ATP supply. While the malate valve maintains baseline energy balance, CEF provides responsive adjustments during acute stress, ensuring efficient and safe photosynthetic operation. Moreover, long-term acclimation processes, such as changes in enzyme expression, adjustments in photosystem stoichiometry, and reprogramming of metabolic pathways including nitrate assimilation and photorespiration, further refine this network, enabling plants to sustain optimal energy balance over extended periods of environmental change [25, 168].

4.2 *Whole-Plant Modulation of NPQ via Source-Sink Balance*

The modulation of NPQ is intricately linked to source-sink balance at whole-plant level [172]. Elevated sink demands, as occurs in young developing leaves and fruits, promote rapid utilization or export of photoassimilates maintaining an effective electron transport rate and minimizing NPQ activation to preserve photochemical efficiency in source leaves. Conversely, compromised sink strength leads to carbohydrate accumulation in source tissues, triggering feedback inhibition on photosynthetic electron flow and enhancing NPQ as a compensatory mechanism to dissipate surplus energy and prevent photoinhibition. Studies in higher plants and algae integrating PAM fluorometry and energy partitioning analyses reveal that metabolic cues, such as carbohydrate accumulation, modulate the kinetics and amplitude of NPQ in real time, safeguarding the photosynthetic machinery from the deleterious effects of overexcitation under stress conditions. This dynamic modulation, where the redox state of the thylakoid membrane acts as a critical regulatory signal, ensures that excess excitation energy is safely dissipated without impairing the photochemical apparatus under conditions of sink limitation [173, 174].

5 Variations and Adaptive Strategies in Natural Environments

The interplay between environmental factors, genetic background, and life strategies shapes the capacity and dynamics of NPQ, enabling plants to optimize photosynthetic efficiency while minimizing photodamage [175–177]. Understanding these diverse mechanisms and their regulation is crucial for further research, potentially leading to improvements in crop resilience and productivity, particularly in the face of changing climatic conditions [178].

The growth environment profoundly influences the capacity for thermal dissipation. Plants grown under fluctuating light conditions, such as those in natural sunlight or shade with sunflecks, can rapidly modulate their light-harvesting efficiency, adjusting NPQ within seconds to minutes. In contrast, plants grown under stable, low-light environments exhibit a limited capacity for thermal dissipation and require hours to adjust their light-harvesting efficiency in response to high light. These differences underscore the importance of exposure to dynamic light conditions in developing efficient photoprotective mechanisms [7, 15].

Genetic factors and life strategies also significantly impact NPQ strategies, as evidenced by the striking contrast between annual and evergreen plants [179, 180]. Rapidly growing annuals, thriving in resource-rich environments, maximize their photosynthetic capacity to fuel rapid growth and reproduction. In contrast, slowly growing evergreens, often facing resource limitations and harsh environmental conditions, prioritize photoprotection through enhanced NPQ mechanisms. These plants reach their maximum photosynthetic capacity at moderate light levels and rely on efficient thermal dissipation to cope with excess light, particularly during periods of high irradiance [180, 181].

Evergreens employ a specific cold acclimation mechanism, referred to as "sustained quenching," where thermal dissipation persists even after extended periods of darkness [180]. This sustained quenching strategy optimizes existing photosynthetic and photoprotective mechanisms, enabling prolonged energy dissipation and enhancing resilience to winter stresses. It involves biochemical adjustments associated with changes in photosynthetic architecture and pigment composition [182, 183]. Notably, evergreens can convert their entire xanthophyll pool to zeaxanthin and antheraxanthin during peak midday irradiance, highlighting their substantial reliance on this mechanism for photoprotection. The sustained and often locked-in NPQ observed in evergreens highlights the importance of this process for plants enduring prolonged periods of abiotic stress [52, 181].

6 Advances in Non-photochemical Quenching Research: Ecological, Agricultural, and Climate Change Implications

Over the past decades, breakthroughs in understanding NPQ have underscored its ecological and agricultural significance, particularly in enhancing crop productivity and predicting plant responses to global climate change. Cutting-edge genetic, biochemical, and systems-level approaches are now being harnessed to optimize NPQ in key crops, with the aim of maximizing photosynthetic efficiency under suboptimal environmental conditions [178, 184–186].

Environmental stressors such as high light intensity, drought, nutrient limitation, and temperature fluctuations critically impact the efficiency and regulation of NPQ, thereby influencing overall plant performance, resilience, and, ultimately, crop yield [15]. These stressors can disrupt the delicate balance between light energy capture and utilization, exacerbating photoinhibition, thus reducing photosynthetic efficiency when NPQ capacity is exceeded. Moreover, photosynthetic adaptations to fluctuating light conditions are not instantaneous, resulting in reduced carbon gain and lower productivity in agricultural crops [187]. Early studies on the transition from dark or shade to sunlight highlighted the challenges in photosynthesis [188]. However, recent studies have emphasized the significance of nonsteady-state responses to fluctuating light intensities as a key strategy to enhance crop photosynthesis and improve resource efficiency [15, 186, 189]. Notably, model simulations indicate that prolonged NPQ relaxation times in crop canopies can lead to a reduction in daily carbon assimilation highlighting the importance of optimizing NPQ dynamics for maximizing photosynthetic efficiency. Enhancing the speed of photosynthetic responses to these fluctuations could increase carbon assimilation by 13–32%, presenting a significant opportunity for yield improvement in major crops [186, 190]. These modifications often involve intricate analyses of the trade-offs between energy dissipation and photosynthetic efficiency, seeking to identify optimal NPQ levels that minimize photodamage while maintaining high rates of carbon fixation. Novel approaches, including successful field trials, have demonstrated the value of this idea, outlining numerous opportunities for improvement, many of which involve transgenic upregulation of key NPQ players [191–194]. Overexpressing PsbS, ZEP, and VDE in tobacco and soybean accelerates NPQ induction and relaxation during sunshade transitions, leading to improved photosynthesis and productivity in field trials [191, 194]. However, similar manipulations in *Arabidopsis* and potato resulted in decreased fresh weight or tuber yield under fluctuating light [195, 196]. These varied outcomes highlight the need for deeper research into the relative importance of these genes and their mechanisms in NPQ formation. As an example, recent studies suggest that varying ZEP concentration is more critical than VDE for accelerating NPQ relaxation [197]. Additionally, advancements in protein engineering through atomistic simulations and gene editing technologies may offer a preferred strategy over transgenic upregulation by enhancing native protein kinetics [186].

However, modifications in NPQ response may disrupt other mechanisms controlling excitation energy distribution or affect source–sink relationships and plastid signaling [195]. An interesting alternative relies on investigating natural variation of elements that improve the speed of photosynthetic adjustments to sunshade transitions to facilitate genomic selection of improved crop varieties [15, 198]. However, given the potential complexity of the relationships between NPQ components and yield, further research is necessary to determine whether strategies aimed at altering NPQ will ultimately benefit crop productivity.

Beyond agricultural applications, understanding the multifaceted role of NPQ in natural ecosystems has far-reaching implications for predicting plant and ecosystem-level responses to ongoing and future climate change scenarios and the development of effective strategies for ecosystem management and conservation [199]. Variations in NPQ capacity among different plant species and functional groups can significantly influence their competitive interactions within diverse ecosystems, particularly under conditions of increased environmental stress [15]. Studies focusing on the acclimation and adaptation of NPQ mechanisms in response to long-term environmental changes, such as elevated atmospheric CO_2 concentrations, altered precipitation patterns, and increased frequency of extreme weather events, are crucial for projecting the future distribution and productivity of plant communities. Moreover, investigating the interactive effects of multiple environmental stressors on NPQ and photosynthetic performance is essential for developing more realistic and accurate predictive models of plant responses to complex environmental change scenarios [200, 201]. In addition, the use of remote sensing techniques to assess NPQ dynamics at landscape scales offers promising avenues for monitoring ecosystem health and productivity as well as for identifying areas that are particularly vulnerable to climate change impacts [199, 202, 203].

In summary, understanding the role of NPQ in mitigating the negative impacts of environmental stress on plant health can inform the development of sustainable land management practices that promote ecosystem resilience and biodiversity conservation [178, 184].

References

1. Geider RJ, Delucia EH, Falkowski PG, Finzi AC, Grime JP, Grace J, Kana TM, La Roche J, Long SP, Osborne BA, Platt T, Prentice IC, Raven JA, Schlesinger WH, Smetacek V, Stuart V, Sathyendranath S, Thomas RB, Vogelmann TC, Williams P, Woodward FI (2001) Primary productivity of planet earth: biological determinants and physical constraints in terrestrial and aquatic habitats. Glob Change Biol 7(8):849–882. https://doi.org/10.1046/j.1365-2486.2001.00448.x
2. Krieger-Liszkay A, Shimakawa G (2022) Regulation of the generation of reactive oxygen species during photosynthetic electron transport. Biochem Soc Trans 50(2):1025–1034. https://doi.org/10.1042/BST20211246
3. Derks A, Schaven K, Bruce D (2015) Diverse mechanisms for photoprotection in photosynthesis. Dynamic regulation of photosystem II excitation in response to rapid environmental change. Biochim Biophys Acta 1847(4–5):468–485. https://doi.org/10.1016/j.bbabio.2015.02.008
4. Li Z, Wakao S, Fischer BB, Niyogi KK (2009) Sensing and responding to excess light. Annu Rev Plant Biol 60(1):239–260. https://doi.org/10.1146/annurev.arplant.58.032806.103844
5. Takahashi S, Badger MR (2011) Photoprotection in plants: a new light on photosystem II damage. Trends Plant Sci 16(1):53–60. https://doi.org/10.1016/j.tplants.2010.10.001
6. Gjindali A, Herrmann HA, Schwartz J-M, Johnson GN, Calzadilla PI (2021) A holistic approach to study photosynthetic acclimation responses of plants to fluctuating light. Front Plant Sci 12:12. https://doi.org/10.3389/fpls.2021.668512
7. Morales A, Kaiser E (2020) Photosynthetic acclimation to fluctuating irradiance in plants. Front Plant Sci 11:268. https://doi.org/10.3389/fpls.2020.00268
8. Hasanuzzaman M, Bhuyan MHMB, Zulfiqar F, Raza A, Mohsin SM, Mahmud JA, Fujita M, Fotopoulos V (2020) Reactive oxygen species and antioxidant defense in plants under abiotic stress: revisiting the crucial role of a universal defense regulator. Antioxidants 9(8):681. https://doi.org/10.3390/antiox9080681
9. Frank HA, Young AJ, Britton G, Cogdell RJ (eds) (1999) The photochemistry of carotenoids. Govindjee, Amesz J, Aro E-M, Barber J, Blankenship RE, Murata N, Ort DR (series eds) Advances in photosynthesis and respiration, vol 8. Springer, Dordrecht. https://doi.org/10.1007/0-306-48209-6.
10. Horton P (1996) Nonphotochemical quenching of chlorophyll fluorescence. In: Jennings RC, Zucchelli G, Ghetti F, Colombetti G (eds) Light as an energy source and information carrier in plant physiology. Springer, Boston, pp 99–111. https://doi.org/10.1007/978-1-4613-0409-8_8
11. Horton P, Ruban A (2005) Molecular design of the photosystem II light-harvesting antenna: photosynthesis and photoprotection. J Exp Bot 56(411):365–373. https://doi.org/10.1093/jxb/eri023
12. Niyogi KK, Truong TB (2013) Evolution of flexible non-photochemical quenching mechanisms that regulate light harvesting in oxygenic photosynthesis. Curr Opin Plant Biol 16(3):307–314. https://doi.org/10.1016/j.pbi.2013.03.011
13. Niyogi KK (1999) Photoprotection revisited: genetic and molecular approaches. Annu Rev Plant Physiol Plant Mol Biol 50:333–359. https://doi.org/10.1146/annurev.arplant.50.1.333
14. Endo T, Uebayashi N, Ishida S, Ikeuchi M, Sato F (2014) Light energy allocation at PSII under field light conditions: how much energy is lost in NPQ-associated dissipation? Plant Physiol Biochem 81:115–120. https://doi.org/10.1016/j.plaphy.2014.03.018

15. Long SP, Taylor SH, Burgess SJ, Carmo-Silva E, Lawson T, De Souza AP, Leonelli L, Wang Y (2022) Into the shadows and back into sunlight: photosynthesis in fluctuating light. Annu Rev Plant Biol 73(1):617–648. https://doi.org/10.1146/annurev-arplant-070221-024745
16. Ruban AV (2016) Nonphotochemical chlorophyll fluorescence quenching: mechanism and effectiveness in protecting plants from photodamage. Plant Physiol 170(4):1903–1916. https://doi.org/10.1104/pp.15.01935
17. Bassi R, Dall'Osto L (2021) Dissipation of light energy absorbed in excess: the molecular mechanisms. Annu Rev Plant Biol 72(1):47–76. https://doi.org/10.1146/annurev-arplant-071720-015522
18. Croce R, Van Amerongen H (2020) Light harvesting in oxygenic photosynthesis: structural biology meets spectroscopy. Science 369(6506):eaay2058. https://doi.org/10.1126/science.aay2058
19. Nicol L, Croce R (2018) Light harvesting in higher plants and green algae. In: van Grondelle R, van Amerongen H, van Stokkum I (eds) Light harvesting in photosynthesis. CRC Press
20. Harbinson J (2018) Chlorophyll fluorescence as a tool for describing the operation and regulation of photosynthesis in vivo. In: Croce R, van Grondelle R, van Amerongen H, van Stokkum I (eds) Light harvesting in photosynthesis. CRC Press
21. Arshad R, Saccon F, Bag P, Biswas A, Calvaruso C, Bhatti AF, Grebe S, Mascoli V, Mahbub M, Muzzopappa F, Polyzois A, Schiphorst C, Sorrentino M, Streckaité S, van Amerongen H, Aro E-M, Bassi R, Boekema EJ, Croce R, Dekker J, van Grondelle R, Jansson S, Kirilovsky D, Kouřil R, Michel S, Mullineaux CW, Panzarová K, Robert B, Ruban AV, van Stokkum I, Wientjes E, Büchel C (2022) A kaleidoscope of photosynthetic antenna proteins and their emerging roles. Plant Physiol 189(3):1204–1219. https://doi.org/10.1093/plphys/kiac175
22. Caffarri S, Tibiletti T, Jennings RC, Santabarbara S (2014) A comparison between plant photosystem I and photosystem II architecture and functioning. Curr Protein Pept Sci 15(4):296–331. https://doi.org/10.2174/1389203715666140327102218
23. Kouřil R, Nosek L, Bartoš J, Boekema EJ, Ilík P (2016) Evolutionary loss of light-harvesting proteins Lhcb6 and Lhcb3 in major land plant groups – break-up of current dogma. New Phytol 210(3):808–814. https://doi.org/10.1111/nph.13947
24. Schiphorst C, Achterberg L, Gómez R, Koehorst R, Bassi R, Van Amerongen H, Dall'Osto L, Wientjes E (2022) The role of light-harvesting complex I in excitation energy transfer from LHCII to photosystem I in Arabidopsis. Plant Physiol 188(4):2241–2252. https://doi.org/10.1093/plphys/kiab579
25. Walker BJ, Kramer DM, Fisher N, Fu X (2020) Flexibility in the energy balancing network of photosynthesis enables safe operation under changing environmental conditions. Plants 9(3):301. https://doi.org/10.3390/plants9030301
26. Goss R, Lepetit B (2015) Biodiversity of NPQ. J Plant Physiol 172:13–32. https://doi.org/10.1016/j.jplph.2014.03.004
27. Müller P, Li X-P, Niyogi KK (2001) Non-photochemical quenching. A response to excess light energy. Plant Physiol 125(4):1558–1566. https://doi.org/10.1104/pp.125.4.1558
28. Schreiber U, Schliwa U, Bilger W (1986) Continuous recording of photochemical and non-photochemical chlorophyll fluorescence quenching with a new type of modulation fluorometer. Photosynth Res 10(1):51–62. https://doi.org/10.1007/BF00024185
29. Baker NR (2008) Chlorophyll fluorescence: a probe of photosynthesis in vivo. Annu Rev Plant Biol 59(1):89–113. https://doi.org/10.1146/annurev.arplant.59.032607.092759
30. Murchie EH, Lawson T (2013) Chlorophyll fluorescence analysis: a guide to good practice and understanding some new applications. J Exp Bot 64(13):3983–3998. https://doi.org/10.1093/jxb/ert208
31. Murchie EH, Ruban AV (2020) Dynamic non-photochemical quenching in plants: from molecular mechanism to productivity. Plant J 101(4):885–896. https://doi.org/10.1111/tpj.14601

32. Nicol L, Nawrocki WJ, Croce R (2019) Disentangling the sites of non-photochemical quenching in vascular plants. Nat Plants 5(11):1177–1183. https://doi.org/10.1038/s41477-019-0526-5
33. Nicol L, Croce R (2021) The PsbS protein and low pH are necessary and sufficient to induce quenching in the light-harvesting complex of plants LHCII. Sci Rep 11(1):7415. https://doi.org/10.1038/s41598-021-86975-9
34. Yang S-J, Wales DJ, Woods EJ, Fleming GR (2024) Design principles for energy transfer in the photosystem II supercomplex from kinetic transition networks. Nat Commun 15(1):8763. https://doi.org/10.1038/s41467-024-53138-z
35. Calzadilla PI, Carvalho FEL, Gomez R, Lima Neto MC, Signorelli S (2022) Assessing photosynthesis in plant systems: a cornerstone to aid in the selection of resistant and productive crops. Environ Exp Bot 201:104950. https://doi.org/10.1016/j.envexpbot.2022.104950
36. Li X-P, Björkman O, Shih C, Grossman AR, Rosenquist M, Jansson S, Niyogi KK (2000) A pigment-binding protein essential for regulation of photosynthetic light harvesting. Nature 403(6768):391–395. https://doi.org/10.1038/35000131
37. Zaks J, Amarnath K, Sylak-Glassman EJ, Fleming GR (2013) Models and measurements of energy-dependent quenching. Photosynth Res 116(2–3):389–409. https://doi.org/10.1007/s11120-013-9857-7
38. Fan M, Li M, Liu Z, Cao P, Pan X, Zhang H, Zhao X, Zhang J, Chang W (2015) Crystal structures of the PsbS protein essential for photoprotection in plants. Nat Struct Mol Biol 22(9):729–735. https://doi.org/10.1038/nsmb.3068
39. Bonente G, Howes BD, Caffarri S, Smulevich G, Bassi R (2008) Interactions between the photosystem II subunit PsbS and xanthophylls studied in vivo and in vitro. J Biol Chem 283(13):8434–8445. https://doi.org/10.1074/jbc.M708291200
40. Li Z, Ahn TK, Avenson TJ, Ballottari M, Cruz JA, Kramer DM, Bassi R, Fleming GR, Keasling JD, Niyogi KK (2009) Lutein accumulation in the absence of zeaxanthin restores nonphotochemical quenching in the *Arabidopsis thaliana Npq1* mutant. Plant Cell 21(6):1798–1812. https://doi.org/10.1105/tpc.109.066571
41. Guardini Z, Gomez RL, Caferri R, Dall'Osto L, Bassi R (2022) Loss of a single chlorophyll in CP29 triggers re-organization of the photosystem II supramolecular assembly. Biochim Biophys Acta Bioenerg 1863(5):148555. https://doi.org/10.1016/j.bbabio.2022.148555
42. Guardini Z, Bressan M, Caferri R, Bassi R, Dall'Osto, L. (2020) Identification of a pigment cluster catalysing fast photoprotective quenching response in CP29. Nat. Plants 6(3):303–313. https://doi.org/10.1038/s41477-020-0612-8
43. Marulanda Valencia W, Pandit A (2024) Photosystem II subunit S (PsbS): a nano regulator of plant photosynthesis. J Mol Biol 436(5):168407. https://doi.org/10.1016/j.jmb.2023.168407
44. Jahns P, Holzwarth AR (2012) The role of the xanthophyll cycle and of lutein in photoprotection of photosystem II. Biochim Biophys Acta Bioenerg 1817(1):182–193. https://doi.org/10.1016/j.bbabio.2011.04.012
45. Li F, Liu C, Streckaite S, Yang C, Xu P, Llansola-Portoles MJ, Ilioaia C, Pascal AA, Croce R, Robert B (2021) A new, unquenched intermediate of LHCII. J Biol Chem 296:296. https://doi.org/10.1016/j.jbc.2021.100322
46. Liu Z, Yan H, Wang K, Kuang T, Zhang J, Gui L, An X, Chang W (2004) Crystal structure of spinach major light-harvesting complex at 2.72 A resolution. Nature 428(6980):287–292. https://doi.org/10.1038/nature02373
47. Dall'Osto L, Lico C, Alric J, Giuliano G, Havaux M, Bassi R (2006) Lutein is needed for efficient chlorophyll triplet quenching in the major LHCII antenna complex of higher plants and effective photoprotection in vivo under strong light. BMC Plant Biol 6(1):32. https://doi.org/10.1186/1471-2229-6-32
48. Lokstein H, Tian L, Polle JEW, DellaPenna D (2002) Xanthophyll biosynthetic mutants of Arabidopsis thaliana: altered nonphotochemical quenching of chlorophyll fluorescence is due to changes in photosystem II antenna size and stability. Biochim Biophys Acta 1553(3):309–319. https://doi.org/10.1016/s0005-2728(02)00184-6

49. Wehner A, Grasses T, Jahns P (2006) De-epoxidation of violaxanthin in the minor antenna proteins of photosystem II, LHCB4, LHCB5, and LHCB6. J Biol Chem 281(31): 21924–21933. https://doi.org/10.1074/jbc.M602915200
50. Croce R, Müller MG, Bassi R, Holzwarth AR (2001) Carotenoid-to-chlorophyll energy transfer in recombinant major light-harvesting complex (LHCII) of higher plants. I. Femtosecond transient absorption measurements. Biophys J 80(2):901–915. https://doi.org/10.1016/S0006-3495(01)76069-9
51. Dall'Osto L, Cazzaniga S, Bressan M, Paleček D, Židek K, Niyogi KK, Fleming GR, Zigmantas D, Bassi R (2017) Two mechanisms for dissipation of excess light in monomeric and trimeric light-harvesting complexes. Nat Plants 3(5):17033. https://doi.org/10.1038/nplants.2017.33
52. Demmig-Adams B, Stewart JJ, López-Pozo M, Polutchko SK, Adams WW (2020) Zeaxanthin, a molecule for photoprotection in many different environments. Molecules 25(24):5825. https://doi.org/10.3390/molecules25245825
53. Horton P, Ruban AV (1992) Regulation of photosystem II. Photosynth Res 34(3):375–385. https://doi.org/10.1007/BF00029812
54. Cignoni E, Lapillo M, Cupellini L, Acosta-Gutiérrez S, Gervasio FL, Mennucci B (2021) A different perspective for nonphotochemical quenching in plant antenna complexes. Nat Commun 12(1):7152. https://doi.org/10.1038/s41467-021-27526-8
55. Sardar S, Caferri R, Camargo FVA, Capaldi S, Ghezzi A, Dall'Osto L, D'Andrea C, Cerullo G, Bassi R (2024) Site-directed mutagenesis of the chlorophyll-binding sites modulates excited-state lifetime and chlorophyll–xanthophyll energy transfer in the monomeric light-harvesting complex CP29. J Phys Chem Lett 15(11):3149–3158. https://doi.org/10.1021/acs.jpclett.3c02900
56. Saccon F, Giovagnetti V, Shukla MK, Ruban AV (2020) Rapid regulation of photosynthetic light harvesting in the absence of minor antenna and reaction centre complexes. J Exp Bot 71(12):3626–3637. https://doi.org/10.1093/jxb/eraa126
57. Bennett DIG, Amarnath K, Park S, Steen CJ, Morris JM, Fleming GR (2019) Models and mechanisms of the rapidly reversible regulation of photosynthetic light harvesting. Open Biol 9(4):190043. https://doi.org/10.1098/rsob.190043
58. Chen L, Rodriguez-Heredia M, Hanke GT, Ruban AV (2025) Distinct features of PsbS essential for mediating plant photoprotection. Plant Commun 6(1):101179. https://doi.org/10.1016/j.xplc.2024.101179
59. Zaks J, Amarnath K, Kramer DM, Niyogi KK, Fleming GR (2012) A kinetic model of rapidly reversible nonphotochemical quenching. Proc Natl Acad Sci 109(39):15757–15762. https://doi.org/10.1073/pnas.1211017109
60. Walla PJ, Holleboom C-P, Fleming GR (2014) Electronic carotenoid-chlorophyll interactions regulating photosynthetic light harvesting of higher plants and green algae. In: Demmig-Adams B, Garab G, Adams III W, Govindjee (eds) Non-photochemical quenching and energy dissipation in plants, algae and cyanobacteria. Springer, Dordrecht, pp 229–243. https://doi.org/10.1007/978-94-017-9032-1_9
61. Ruban AV, Berera R, Ilioaia C, Van Stokkum IHM, Kennis JTM, Pascal AA, Van Amerongen H, Robert B, Horton P, Van Grondelle R (2007) Identification of a mechanism of photoprotective energy dissipation in higher plants. Nature 450(7169):575–578. https://doi.org/10.1038/nature06262
62. Fox KF, Balevičius V, Chmeliov J, Valkunas L, Ruban AV, Duffy CDP (2017) The carotenoid pathway: what is important for excitation quenching in plant antenna complexes? Phys Chem Chem Phys 19(34):22957–22968. https://doi.org/10.1039/C7CP03535G
63. Ruban AV, Wilson S (2021) The mechanism of non-photochemical quenching in plants: localization and driving forces. Plant Cell Physiol 62(7):1063–1072. https://doi.org/10.1093/pcp/pcaa155
64. Ahn TK, Avenson TJ, Ballottari M, Cheng Y-C, Niyogi KK, Bassi R, Fleming GR (2008) Architecture of a charge-transfer state regulating light harvesting in a plant antenna protein. Science 320(5877):794–797. https://doi.org/10.1126/science.1154800

65. Avenson TJ, Ahn TK, Zigmantas D, Niyogi KK, Li Z, Ballottari M, Bassi R, Fleming GR (2008) Zeaxanthin radical cation formation in minor light-harvesting complexes of higher plant antenna. J Biol Chem 283(6):3550–3558. https://doi.org/10.1074/jbc.M705645200
66. Holt NE, Zigmantas D, Valkunas L, Li X-P, Niyogi KK, Fleming GR (2005) Carotenoid cation formation and the regulation of photosynthetic light harvesting. Science 307(5708): 433–436. https://doi.org/10.1126/science.1105833
67. Amarie S, Wilk L, Barros T, Kühlbrandt W, Dreuw A, Wachtveitl J (2009) Properties of zeaxanthin and its radical cation bound to the minor light-harvesting complexes CP24, CP26 and CP29. Biochim Biophys Acta Bioenerg 1787(6):747–752. https://doi.org/10.1016/j.bbabio.2009.02.006
68. Holzwarth AR, Miloslavina Y, Nilkens M, Jahns P (2009) Identification of two quenching sites active in the regulation of photosynthetic light-harvesting studied by time-resolved fluorescence. Chem Phys Lett 483(4):262–267. https://doi.org/10.1016/j.cplett.2009.10.085
69. Miloslavina Y, Wehner A, Lambrev PH, Wientjes E, Reus M, Garab G, Croce R, Holzwarth AR (2008) Far-red fluorescence: a direct spectroscopic marker for LHCII oligomer formation in non-photochemical quenching. FEBS Lett 582(25–26):3625–3631. https://doi.org/10.1016/j.febslet.2008.09.044
70. Lapillo M, Cignoni E, Cupellini L, Mennucci B (2020) The energy transfer model of nonphotochemical quenching: lessons from the minor CP29 antenna complex of plants. Biochim Biophys Acta Bioenerg 1861(11):148282. https://doi.org/10.1016/j.bbabio.2020.148282
71. Bonente G, Ballottari M, Truong TB, Morosinotto T, Ahn TK, Fleming GR, Niyogi KK, Bassi R (2011) Analysis of LhcSR3, a protein essential for feedback de-excitation in the green alga Chlamydomonas reinhardtii. PLoS Biol 9(1):e1000577. https://doi.org/10.1371/journal.pbio.1000577
72. Bode S, Quentmeier CC, Liao P-N, Hafi N, Barros T, Wilk L, Bittner F, Walla PJ (2009) On the regulation of photosynthesis by excitonic interactions between carotenoids and chlorophylls. Proc Natl Acad Sci 106(30):12311–12316. https://doi.org/10.1073/pnas.0903536106
73. Leonardo C, Yang S-J, Orcutt K, Iwai M, Arsenault EA, Fleming GR (2024) Bidirectional energy flow in the photosystem II supercomplex. J Phys Chem B 128(33):7941–7953. https://doi.org/10.1021/acs.jpcb.4c02508
74. Horton P, Ruban AV, Rees D, Pascal AA, Noctor G, Young AJ (1991) Control of the light-harvesting function of chloroplast membranes by aggregation of the LHCII chlorophyll-protein complex. FEBS Lett 292(1–2):1–4. https://doi.org/10.1016/0014-5793(91)80819-o
75. Ilioaia C, Johnson MP, Horton P, Ruban AV (2008) Induction of efficient energy dissipation in the isolated light-harvesting complex of photosystem II in the absence of protein aggregation. J Biol Chem 283(43):29505–29512. https://doi.org/10.1074/jbc.M802438200
76. Malý P, Gruber JM, van Grondelle R, Mančal T (2016) Single molecule spectroscopy of monomeric LHCII: experiment and theory. Sci Rep 6(1):26230. https://doi.org/10.1038/srep26230
77. Johnson MP, Goral TK, Duffy CDP, Brain APR, Mullineaux CW, Ruban AV (2011) Photoprotective energy dissipation involves the reorganization of photosystem II light-harvesting complexes in the grana membranes of spinach chloroplasts. Plant Cell 23(4): 1468–1479. https://doi.org/10.1105/tpc.110.081646
78. Akhtar P, Görföl F, Garab G, Lambrev PH (2019) Dependence of chlorophyll fluorescence quenching on the lipid-to-protein ratio in reconstituted light-harvesting complex II membranes containing lipid labels. Chem Phys 522:242–248. https://doi.org/10.1016/j.chemphys.2019.03.012
79. Tietz S, Leuenberger M, Höhner R, Olson AH, Fleming GR, Kirchhoff H (2020) A proteoliposome-based system reveals how lipids control photosynthetic light harvesting. J Biol Chem 295(7):1857–1866. https://doi.org/10.1074/jbc.RA119.011707
80. Daskalakis V, Papadatos S, Kleinekathöfer U (2019) Fine tuning of the photosystem II major antenna mobility within the thylakoid membrane of higher plants. Biochim Biophys Acta Biomembr 1861(12):183059. https://doi.org/10.1016/j.bbamem.2019.183059

81. Navakoudis E, Stergiannakos T, Daskalakis V (2023) A perspective on the major light-harvesting complex dynamics under the effect of pH, salts, and the photoprotective PsbS protein. Photosynth Res 156(1):163–177. https://doi.org/10.1007/s11120-022-00935-6
82. Welc R, Luchowski R, Kluczyk D, Zubik-Duda M, Grudzinski W, Maksim M, Reszczynska E, Sowinski K, Mazur R, Nosalewicz A, Gruszecki WI (2021) Mechanisms shaping the synergism of zeaxanthin and PsbS in photoprotective energy dissipation in the photosynthetic apparatus of plants. Plant J 107(2):418–433. https://doi.org/10.1111/tpj.15297
83. Lambrev PH, Nilkens M, Miloslavina Y, Jahns P, Holzwarth AR (2010) Kinetic and spectral resolution of multiple nonphotochemical quenching components in Arabidopsis leaves. Plant Physiol 152(3):1611–1624. https://doi.org/10.1104/pp.109.148213
84. Nilkens M, Kress E, Lambrev P, Miloslavina Y, Müller M, Holzwarth AR, Jahns P (2010) Identification of a slowly inducible zeaxanthin-dependent component of non-photochemical quenching of chlorophyll fluorescence generated under steady-state conditions in Arabidopsis. Biochim Biophys Acta 1797(4):466–475. https://doi.org/10.1016/j.bbabio.2010.01.001
85. Townsend AJ, Saccon F, Giovagnetti V, Wilson S, Ungerer P, Ruban AV (2018) The causes of altered chlorophyll fluorescence quenching induction in the Arabidopsis mutant lacking all minor antenna complexes. Biochim Biophys Acta Bioenerg 1859(9):666–675. https://doi.org/10.1016/j.bbabio.2018.03.005
86. Rochaix J-D (2014) Regulation and dynamics of the light-harvesting system. Annu Rev Plant Biol 65:287–309. https://doi.org/10.1146/annurev-arplant-050213-040226
87. Allen JF (2003) State transitions – a question of balance. Science 299(5612):1530–1532. https://doi.org/10.1126/science.1082833
88. Goldschmidt-Clermont M, Bassi R (2015) Sharing light between two photosystems: mechanism of state transitions. Curr Opin Plant Biol 25:71–78. https://doi.org/10.1016/j.pbi.2015.04.009
89. Bellafiore S, Barneche F, Peltier G, Rochaix J-D (2005) State transitions and light adaptation require chloroplast thylakoid protein kinase STN7. Nature 433(7028):892–895. https://doi.org/10.1038/nature03286
90. Bonardi V, Pesaresi P, Becker T, Schleiff E, Wagner R, Pfannschmidt T, Jahns P, Leister D (2005) Photosystem II core phosphorylation and photosynthetic acclimation require two different protein kinases. Nature 437(7062):1179–1182. https://doi.org/10.1038/nature04016
91. Pribil M, Pesaresi P, Hertle A, Barbato R, Leister D (2010) Role of plastid protein phosphatase TAP38 in LHCII dephosphorylation and thylakoid electron flow. PLoS Biol 8(1):e1000288. https://doi.org/10.1371/journal.pbio.1000288
92. Shapiguzov A, Ingelsson B, Samol I, Andres C, Kessler F, Rochaix J-D, Vener AV, Goldschmidt-Clermont M (2010) The PPH1 phosphatase is specifically involved in LHCII dephosphorylation and state transitions in Arabidopsis. Proc Natl Acad Sci USA 107(10): 4782–4787. https://doi.org/10.1073/pnas.0913810107
93. Pinnola A (2019) The rise and fall of light-harvesting complex stress-related proteins as photoprotection agents during evolution. J Exp Bot 70(20):5527–5535. https://doi.org/10.1093/jxb/erz317
94. Gerotto C, Morosinotto T (2013) Evolution of photoprotection mechanisms upon land colonization: evidence of PSBS-dependent NPQ in late Streptophyte algae. Physiol Plant 149(4): 583–598. https://doi.org/10.1111/ppl.12070
95. Alboresi A, Gerotto C, Giacometti GM, Bassi R, Morosinotto T (2010) Physcomitrella patens mutants affected on heat dissipation clarify the evolution of photoprotection mechanisms upon land colonization. Proc Natl Acad Sci 107(24):11128–11133. https://doi.org/10.1073/pnas.1002873107
96. Gerotto C, Alboresi A, Giacometti GM, Bassi R, Morosinotto T (2012) Coexistence of plant and algal energy dissipation mechanisms in the moss Physcomitrella patens. New Phytol 196(3):763–773. https://doi.org/10.1111/j.1469-8137.2012.04345.x
97. Gerotto C, Alboresi A, Giacometti GM, Bassi R, Morosinotto T (2011) Role of PSBS and LHCSR in Physcomitrella patens acclimation to high light and low temperature. Plant Cell Environ 34(6):922–932. https://doi.org/10.1111/j.1365-3040.2011.02294.x

98. Bonente G, Passarini F, Cazzaniga S, Mancone C, Buia MC, Tripodi M, Bassi R, Caffarri S (2008) The occurrence of the psbS gene product in Chlamydomonas reinhardtii and in other photosynthetic organisms and its correlation with energy quenching. Photochem Photobiol 84(6):1359–1370. https://doi.org/10.1111/j.1751-1097.2008.00456.x
99. Tian L, Nawrocki WJ, Liu X, Polukhina I, van Stokkum IHM, Croce R (2019) pH dependence, kinetics and light-harvesting regulation of nonphotochemical quenching in Chlamydomonas. Proc Natl Acad Sci 116(17):8320–8325. https://doi.org/10.1073/pnas.1817796116
100. Pinnola A, Bassi R (2018) Molecular mechanisms involved in plant photoprotection. Biochem Soc Trans 46(2):467–482. https://doi.org/10.1042/BST20170307
101. Pinnola A, Staleva-Musto H, Capaldi S, Ballottari M, Bassi R, Polívka T (2016) Electron transfer between carotenoid and chlorophyll contributes to quenching in the LHCSR1 protein from Physcomitrella patens. Biochim Biophys Acta 1857(12):1870–1878. https://doi.org/10.1016/j.bbabio.2016.09.001
102. Quaas T, Berteotti S, Ballottari M, Flieger K, Bassi R, Wilhelm C, Goss R (2015) Non-photochemical quenching and xanthophyll cycle activities in six green algal species suggest mechanistic differences in the process of excess energy dissipation. J Plant Physiol 172:92–103. https://doi.org/10.1016/j.jplph.2014.07.023
103. Blommaert L, Chafai L, Bailleul B (2021) The fine-tuning of NPQ in diatoms relies on the regulation of both xanthophyll cycle enzymes. Sci Rep 11(1):12750. https://doi.org/10.1038/s41598-021-91483-x
104. Pinnola A, Cazzaniga S, Alboresi A, Nevo R, Levin-Zaidman S, Reich Z, Bassi R (2015) Light-harvesting complex stress-related proteins catalyze excess energy dissipation in both photosystems of Physcomitrella patens. Plant Cell 27(11):3213–3227. https://doi.org/10.1105/tpc.15.00443
105. Nawrocki WJ, Santabarbara S, Mosebach L, Wollman F-A, Rappaport F (2016) State transitions redistribute rather than dissipate energy between the two photosystems in Chlamydomonas. Nat Plants 2(4):16031. https://doi.org/10.1038/nplants.2016.31
106. Aro E-M, Virgin I, Andersson B, Photoinhibition of photosystem II. Inactivation, protein damage and turnover (1993) Biochim Biophys Acta Bioenerg 1143(2):113–134. https://doi.org/10.1016/0005-2728(93)90134-2
107. Krause GH, Weis E (1991) Chlorophyll fluorescence and photosynthesis: the basics. Annu Rev Plant Biol 42:313–349. https://doi.org/10.1146/annurev.pp.42.060191.001525
108. Chow WS, Lee H-Y, He J, Hendrickson L, Hong Y-N, Matsubara S (2005) Photoinactivation of photosystem II in leaves. Photosynth Res 84(1–3):35–41. https://doi.org/10.1007/s11120-005-0410-1
109. Tikkanen M, Aro E-M (2012) Thylakoid protein phosphorylation in dynamic regulation of photosystem II in higher plants. Biochim Biophys Acta Bioenerg 1817(1):232–238. https://doi.org/10.1016/j.bbabio.2011.05.005
110. Nishiyama Y, Murata N (2014) Revised scheme for the mechanism of photoinhibition and its application to enhance the abiotic stress tolerance of the photosynthetic machinery. Appl Microbiol Biotechnol 98(21):8777–8796. https://doi.org/10.1007/s00253-014-6020-0
111. Järvi S, Suorsa M, Aro E-M (2015) Photosystem II repair in plant chloroplasts – regulation, assisting proteins and shared components with photosystem II biogenesis. Biochim Biophys Acta Bioenerg 1847(9):900–909. https://doi.org/10.1016/j.bbabio.2015.01.006
112. Tikkanen M, Mekala NR, Aro E-M (2014) Photosystem II photoinhibition-repair cycle protects photosystem I from irreversible damage. Biochim Biophys Acta Bioenerg 1837(1):210–215. https://doi.org/10.1016/j.bbabio.2013.10.001
113. Dall'Osto L, Caffarri S, Bassi R (2005) A mechanism of nonphotochemical energy dissipation, independent from PsbS, revealed by a conformational change in the antenna protein CP26. Plant Cell 17(4):1217–1232. https://doi.org/10.1105/tpc.104.030601
114. Kress E, Jahns P (2017) The dynamics of energy dissipation and xanthophyll conversion in Arabidopsis indicate an indirect photoprotective role of zeaxanthin in slowly inducible and

relaxing components of non-photochemical quenching of excitation energy. Front Plant Sci 8: 2094. https://doi.org/10.3389/fpls.2017.02094

115. Brooks MD, Sylak-Glassman EJ, Fleming GR, Niyogi KK (2013) A thioredoxin-like/β-propeller protein maintains the efficiency of light harvesting in Arabidopsis. Proc Natl Acad Sci 110(29):E2733–E2740. https://doi.org/10.1073/pnas.1305443110
116. Bru P, Steen CJ, Park S, Amstutz CL, Sylak-Glassman EJ, Lam L, Fekete A, Mueller MJ, Longoni F, Fleming GR, Niyogi KK, Malnoë A (2022) The major trimeric antenna complexes serve as a site for qH-energy dissipation in plants. J Biol Chem 298(11):102519. https://doi.org/10.1016/j.jbc.2022.102519
117. Malnoë A (2018) Photoinhibition or photoprotection of photosynthesis? Update on the (newly termed) sustained quenching component qH. Environ Exp Bot 154:123–133. https://doi.org/10.1016/j.envexpbot.2018.05.005
118. Cazzaniga S, Dall'Osto L, Kong S-G, Wada M, Bassi R (2013) Interaction between avoidance of photon absorption, excess energy dissipation and zeaxanthin synthesis against photooxidative stress in Arabidopsis. Plant J 76(4):568–579. https://doi.org/10.1111/tpj.12314
119. Dall'Osto L, Cazzaniga S, Wada M, Bassi R (2014) On the origin of a slowly reversible fluorescence decay component in the Arabidopsis Npq4 mutant. Philos Trans R Soc Lond B Biol Sci 369(1640):20130221. https://doi.org/10.1098/rstb.2013.0221
120. Wada M, Kagawa T, Sato Y (2003) Chloroplast movement. Annu Rev Plant Biol 54:455–468. https://doi.org/10.1146/annurev.arplant.54.031902.135023
121. Kasahara M, Kagawa T, Oikawa K, Suetsugu N, Miyao M, Wada M (2002) Chloroplast avoidance movement reduces photodamage in plants. Nature 420(6917):829–832. https://doi.org/10.1038/nature01213
122. Kalaji HM, Schansker G, Brestic M, Bussotti F, Calatayud A, Ferroni L, Goltsev V, Guidi L, Jajoo A, Li P, Losciale P, Mishra VK, Misra AN, Nebauer SG, Pancaldi S, Penella C, Pollastrini M, Suresh K, Tambussi E, Yanniccari M, Zivcak M, Cetner MD, Samborska IA, Stirbet A, Olsovska K, Kunderlikova K, Shelonzek H, Rusinowski S, Bąba W (2017) Frequently asked questions about chlorophyll fluorescence, the sequel. Photosynth Res 132(1):13–66. https://doi.org/10.1007/s11120-016-0318-y
123. Kalaji HM, Schansker G, Ladle RJ, Goltsev V, Bosa K, Allakhverdiev SI, Brestic M, Bussotti F, Calatayud A, Dąbrowski P, Elsheery NI, Ferroni L, Guidi L, Hogewoning SW, Jajoo A, Misra AN, Nebauer SG, Pancaldi S, Penella C, Poli D, Pollastrini M, Romanowska-Duda ZB, Rutkowska B, Serôdio J, Suresh K, Szulc W, Tambussi E, Yanniccari M, Zivcak M (2014) Frequently asked questions about in vivo chlorophyll fluorescence: practical issues. Photosynth Res 122(2):121–158. https://doi.org/10.1007/s11120-014-0024-6
124. Maxwell K, Johnson GN (2000) Chlorophyll fluorescence – a practical guide. J Exp Bot 51(345):659–668. https://doi.org/10.1093/jexbot/51.345.659
125. Oxborough K, Baker NR (1997) Resolving chlorophyll a fluorescence images of photosynthetic efficiency into photochemical and non-photochemical components – calculation of qP and Fv-/Fm-; without measuring Fo–. Photosynth Res 54(2):135–142. https://doi.org/10.1023/A:1005936823310
126. Bilger W, Björkman O (1990) Role of the xanthophyll cycle in photoprotection elucidated by measurements of light-induced absorbance changes, fluorescence and photosynthesis in leaves of Hedera canariensis. Photosynth Res 25(3):173–185. https://doi.org/10.1007/BF00033159
127. Tietz S, Hall CC, Cruz JA, Kramer DM (2017) $NPQ_{(T)}$: a chlorophyll fluorescence parameter for rapid estimation and imaging of non-photochemical quenching of excitons in photosystem-II-associated antenna complexes. Plant Cell Environ 40(8):1243–1255. https://doi.org/10.1111/pce.12924
128. Lauterberg M, Tschiersch H, Zhao Y, Kuhlmann M, Mücke I, Papa R, Bitocchi E, Neumann K (2024) Implementation of theoretical non-photochemical quenching (NPQ(T)) to investigate NPQ of chickpea under drought stress with high-throughput phenotyping. Sci Rep 14(1): 13970. https://doi.org/10.1038/s41598-024-63372-6

129. Liu S, Xiong Z, Zhang Z, Wei Y, Xiong D, Wang F, Huang J (2023) Exploration of chlorophyll fluorescence characteristics gene regulatory in rice (Oryza Sativa L.): a genome-wide association study. Front Plant Sci 14:1234866. https://doi.org/10.3389/fpls.2023.1234866
130. Sahay S, Grzybowski M, Schnable JC, Głowacka K (2024) Genotype-specific nonphotochemical quenching responses to nitrogen deficit are linked to chlorophyll a to b ratios. J Plant Physiol 297:154261. https://doi.org/10.1016/j.jplph.2024.154261
131. Kuhlgert S, Austic G, Zegarac R, Osei-Bonsu I, Hoh D, Chilvers MI, Roth MG, Bi K, TerAvest D, Weebadde P, Kramer DM (2016) MultispeQ beta: a tool for large-scale plant phenotyping connected to the open PhotosynQ network. R Soc Open Sci 3(10):160592. https://doi.org/10.1098/rsos.160592
132. Ruban AV, Murchie EH (2012) Assessing the photoprotective effectiveness of non-photochemical chlorophyll fluorescence quenching: a new approach. Biochim Biophys Acta Bioenerg 1817(7):977–982. https://doi.org/10.1016/j.bbabio.2012.03.026
133. Ruban AV (1730) Quantifying the efficiency of photoprotection. Philos Trans R Soc B Biol Sci 2017(372):20160393. https://doi.org/10.1098/rstb.2016.0393
134. Ruban AV, Belgio E (1640) The relationship between maximum tolerated light intensity and photoprotective energy dissipation in the photosynthetic antenna: chloroplast gains and losses. Philos Trans R Soc B Biol Sci 2014(369):20130222. https://doi.org/10.1098/rstb.2013.0222
135. Carvalho FEL, Ware MA, Ruban AV (2015) Quantifying the dynamics of light tolerance in Rabidopsis plants during ontogenesis. Plant Cell Environ 38(12):2603–2617. https://doi.org/10.1111/pce.12574
136. Foo CC, Burgess AJ, Retkute R, Tree-Intong P, Ruban AV, Murchie EH (2020) Photoprotective energy dissipation is greater in the lower, not the upper, regions of a rice canopy: a 3D analysis. J Exp Bot 71(22):7382–7392. https://doi.org/10.1093/jxb/eraa411
137. Ware MA, Belgio E, Ruban AV (2015) Comparison of the protective effectiveness of NPQ in Arabidopsis plants deficient in PsbS protein and zeaxanthin. J Exp Bot 66(5):1259–1270. https://doi.org/10.1093/jxb/eru477
138. Lazár D (2015) Parameters of photosynthetic energy partitioning. J Plant Physiol 175:131–147. https://doi.org/10.1016/j.jplph.2014.10.021
139. Demmig-Adams B, Adams III WW, Barker DH, Logan BA, Bowling DR, Verhoeven AS (1996) Using chlorophyll fluorescence to assess the fraction of absorbed light allocated to thermal dissipation of excess excitation. Physiol Plant 98(2):253–264. https://doi.org/10.1034/j.1399-3054.1996.980206.x
140. Hendrickson L, Förster B, Pogson BJ, Chow WS (2005) A simple chlorophyll fluorescence parameter that correlates with the rate coefficient of photoinactivation of photosystem II. Photosynth Res 84(1):43–49. https://doi.org/10.1007/s11120-004-6430-4
141. Hendrickson L, Furbank RT, Chow WS (2004) A simple alternative approach to assessing the fate of absorbed light energy using chlorophyll fluorescence. Photosynth Res 82(1):73–81. https://doi.org/10.1023/B:PRES.0000040446.87305.f4
142. Kramer DM, Johnson G, Kiirats O, Edwards GE (2004) New fluorescence parameters for the determination of Q_A redox state and excitation energy fluxes. Photosynth Res 79(2):209–218. https://doi.org/10.1023/B:PRES.0000015391.99477.0d
143. Ahn TK, Avenson TJ, Peers G, Li Z, Dall'Osto L, Bassi R, Niyogi KK, Fleming GR (2009) Investigating energy partitioning during photosynthesis using an expanded quantum yield convention. Chem Phys 357(1–3):151–158. https://doi.org/10.1016/j.chemphys.2008.12.003
144. Kasajima I, Takahara K, Kawai-Yamada M, Uchimiya H (2009) Estimation of the relative sizes of rate constants for chlorophyll de-excitation processes through comparison of inverse fluorescence intensities. Plant Cell Physiol 50(9):1600–1616. https://doi.org/10.1093/pcp/pcp102
145. Guadagno CR, Virzo De Santo A, D'Ambrosio N (2010) A revised energy partitioning approach to assess the yields of non-photochemical quenching components. Biochim Biophys Acta Bioenerg 1797(5):525–530. https://doi.org/10.1016/j.bbabio.2010.01.016

146. Kornyeyev D, Guadagno CR, D'Ambrosio N (2013) Assessment of energy partitioning in PSII complexes using chlorophyll fluorescence: reviewing the different approaches toward the definition of a unified method. Photosynthetica 51(3):477–480. https://doi.org/10.1007/s11099-013-0027-0
147. Porcar-Castell A, Pfündel E, Korhonen J, Juurola E (2008) A new monitoring PAM fluorometer (MONI-PAM) to study the short- and long-term acclimation of photosystem II in field conditions. Photosynth Res 96:173–179. https://doi.org/10.1007/s11120-008-9292-3
148. Logan BA, Demmig-Adams B, Adams WW, Bilger W (2014) Context, quantification, and measurement guide for non-photochemical quenching of chlorophyll fluorescence. In: Demmig-Adams B, Garab G, Adams III W, Govindjee (eds) Non-photochemical quenching and energy dissipation in plants, algae and cyanobacteria. Advances in photosynthesis and respiration, vol 40. Springer, Dordrecht, pp 187–201. https://doi.org/10.1007/978-94-017-9032-1_7
149. Logan BA, Adams WW, Demmig-Adams B (2007) Viewpoint: avoiding common pitfalls of chlorophyll fluorescence analysis under field conditions. Funct Plant Biol 34(9):853. https://doi.org/10.1071/FP07113
150. Markgraf T, Berry J (1990) Measurement of photochemical and non-photochemical quenching: correction for turnover of PS2 during steady-state photosynthesis. In: Baltscheffsky M (ed) Current research in photosynthesis: proceedings of the VIIIth international conference on photosynthesis, Stockholm, Sweden, Aug 6–11, 1989. Springer, Dordrecht, pp 3073–3076. https://doi.org/10.1007/978-94-009-0511-5_694
151. Kramer DM, Crofts AR (2004) Control and measurement of photosynthetic electron transport in vivo. In: Baker NR (ed) Photosynthesis and the environment. Advances in photosynthesis and respiration, vol 5. Kluwer Academic Publishers, Dordrecht, pp 25–66. https://doi.org/10.1007/0-306-48135-9_2
152. Schreiber U (2004) Pulse-amplitude-modulation (PAM) fluorometry and saturation pulse method: an overview. In: Papageorgiou GC, Govindjee (eds) Chlorophyll a fluorescence: a signature of photosynthesis. Springer, Dordrecht, pp 279–319. https://doi.org/10.1007/978-1-4020-3218-9_11
153. Loriaux SD, Avenson TJ, Welles JM, Mcdermitt DK, Eckles RD, Riensche B, Genty B (2013) Closing in on maximum yield of chlorophyll fluorescence using a single multiphase flash of sub-saturating intensity. Plant Cell Environ 36(10):1755–1770. https://doi.org/10.1111/pce.12115
154. Pfündel EE (2009) Deriving room temperature excitation spectra for photosystem I and photosystem II fluorescence in intact leaves from the dependence of FV/FM on excitation wavelength. Photosynth Res 100(3):163–177. https://doi.org/10.1007/s11120-009-9453-z
155. Pfündel EE, Klughammer C, Meister A, Cerovic ZG (2013) Deriving fluorometer-specific values of relative PSI fluorescence intensity from quenching of F0 fluorescence in leaves of Arabidopsis thaliana and Zea mays. Photosynth Res 114(3):189–206. https://doi.org/10.1007/s11120-012-9788-8
156. Terashima I, Oguchi R, Atsuzawa K, Kaneko Y, Kono M (2025) Excitation spillover from PSII to PSI measured in leaves at 77 K. Plant Cell Physiol 66:pcaf002. https://doi.org/10.1093/pcp/pcaf002
157. Ueno Y, Aikawa S, Kondo A, Akimoto S (2016) Energy transfer in cyanobacteria and red algae: confirmation of spillover in intact megacomplexes of phycobilisome and both photosystems. J Phys Chem Lett 7(18):3567–3571. https://doi.org/10.1021/acs.jpclett.6b01609
158. Ramakers LAI, Harbinson J, Wientjes E, Van Amerongen H (2025) Unravelling the different components of nonphotochemical quenching using a novel analytical pipeline. New Phytol 245(2):625–636. https://doi.org/10.1111/nph.20271
159. Lamb JJ, Røkke G, Hohmann-Marriott MF (2018) Chlorophyll fluorescence emission spectroscopy of oxygenic organisms at 77 K. Photosynthetica 56(1):105–124. https://doi.org/10.1007/s11099-018-0791-y

160. Chukhutsina VU, Holzwarth AR, Croce R (2019) Time-resolved fluorescence measurements on leaves: principles and recent developments. Photosynth Res 140(3):355–369. https://doi.org/10.1007/s11120-018-0607-8
161. Kolber ZS, Prášil O, Falkowski PG (1998) Measurements of variable chlorophyll fluorescence using fast repetition rate techniques: defining methodology and experimental protocols. Biochim Biophys Acta Bioenerg 1367(1):88–106. https://doi.org/10.1016/S0005-2728(98)00135-2
162. Keller B, Vass I, Matsubara S, Paul K, Jedmowski C, Pieruschka R, Nedbal L, Rascher U, Muller O (2019) Maximum fluorescence and electron transport kinetics determined by light-induced fluorescence transients (LIFT) for photosynthesis phenotyping. Photosynth Res 140(2):221–233. https://doi.org/10.1007/s11120-018-0594-9
163. Kolber Z, Klimov D, Ananyev G, Rascher U, Berry J, Osmond B (2005) Measuring photosynthetic parameters at a distance: laser induced fluorescence transient (LIFT) method for remote measurements of photosynthesis in terrestrial vegetation. Photosynth Res 84(1): 121–129. https://doi.org/10.1007/s11120-005-5092-1
164. Fu P, Montes CM, Siebers MH, Gomez-Casanovas N, McGrath JM, Ainsworth EA, Bernacchi CJ (2022) Advances in field-based high-throughput photosynthetic phenotyping. J Exp Bot 73(10):3157–3172. https://doi.org/10.1093/jxb/erac077
165. Zendonadi Dos Santos N, Piepho H, Condorelli GE, Licieri Groli E, Newcomb M, Ward R, Tuberosa R, Maccaferri M, Fiorani F, Rascher U, Muller O (2021) High-throughput field phenotyping reveals genetic variation in photosynthetic traits in durum wheat under drought. Plant Cell Environ 44(9):2858–2878. https://doi.org/10.1111/pce.14136
166. Porcar-Castell A, Malenovský Z, Magney T, Van Wittenberghe S, Fernández-Marín B, Maignan F, Zhang Y, Maseyk K, Atherton J, Albert LP, Robson TM, Zhao F, Garcia-Plazaola J-I, Ensminger I, Rajewicz PA, Grebe S, Tikkanen M, Kellner JR, Ihalainen JA, Rascher U, Logan B (2021) Chlorophyll a fluorescence illuminates a path connecting plant molecular biology to earth-system science. Nat Plants 7(8):998–1009. https://doi.org/10.1038/s41477-021-00980-4
167. Torrado B, Pannunzio B, Malacrida L, Digman MA (2024) Fluorescence lifetime imaging microscopy. Nat Rev Methods Primer 4(1):80. https://doi.org/10.1038/s43586-024-00358-8
168. Shikanai T (2024) Molecular genetic dissection of the regulatory network of proton motive force in chloroplasts. Plant Cell Physiol 65(4):537–550. https://doi.org/10.1093/pcp/pcad157
169. Fisher N, Bricker TM, Kramer DM (2019) Regulation of photosynthetic cyclic electron flow pathways by adenylate status in higher plant chloroplasts. Biochimica et Biophysica Acta (BBA) - Bioenergetics 1860(11):148081. https://doi.org/10.1016/j.bbabio.2019.148081
170. Kanazawa A, Ostendorf E, Kohzuma K, Hoh D, Strand DD, Sato-Cruz M, Savage L, Cruz JA, Fisher N, Froehlich JE, Kramer DM (2017) Chloroplast ATP synthase modulation of the thylakoid proton motive force: implications for photosystem I and photosystem II photoprotection. Front Plant Sci 8:8. https://doi.org/10.3389/fpls.2017.00719
171. Scheibe R (2019) Maintaining homeostasis by controlled alternatives for energy distribution in plant cells under changing conditions of supply and demand. Photosynth Res 139(1):81–91. https://doi.org/10.1007/s11120-018-0583-z
172. Paul MJ, Foyer CH (2001) Sink regulation of photosynthesis. J Exp Bot 52(360):1383–1400. https://doi.org/10.1093/jexbot/52.360.1383
173. Adams WW, Muller O, Cohu CM, Demmig-Adams B (2014) Photosystem II efficiency and non-photochemical fluorescence quenching in the context of source-sink balance. In: Demmig-Adams B, Garab G, Adams III W, Govindjee (eds) Non-photochemical quenching and energy dissipation in plants, algae and cyanobacteria. Springer, Dordrecht, pp 503–529. https://doi.org/10.1007/978-94-017-9032-1_23
174. Rosado-Souza L, Yokoyama R, Sonnewald U, Fernie AR (2023) Understanding source–sink interactions: progress in model plants and translational research to crops. Mol Plant 16(1): 96–121. https://doi.org/10.1016/j.molp.2022.11.015

175. Demmig-Adams B, Stewart JJ, Adams WW (2019) Less photoprotection can be good in some genetic and environmental contexts. Biochem J 476(14):2017–2029. https://doi.org/10.1042/BCJ20190328
176. Kaiser E, Morales A, Harbinson J, Kromdijk J, Heuvelink E, Marcelis LFM (2015) Dynamic photosynthesis in different environmental conditions. J Exp Bot 66(9):2415–2426. https://doi.org/10.1093/jxb/eru406
177. Sello S, Meneghesso A, Alboresi A, Baldan B, Morosinotto T (2019) Plant biodiversity and regulation of photosynthesis in the natural environment. Planta 249(4):1217–1228. https://doi.org/10.1007/s00425-018-03077-z
178. Pesaresi P, Bono P, Corn S, Crosatti C, Daniotti S, Jensen JD, Frébort I, Groli E, Halpin C, Hansson M, Hensel G, Horner DS, Houston K, Jahoor A, Klíma M, Kollist H, Lacoste C, Laidoudi B, Larocca S, Marè C, Moigne NL, Mizzotti C, Morosinotto T, Oldach K, Rossini L, Raubach S, Sanchez-Garcia M, Shaw PD, Sonnier R, Tondelli A, Waugh R, Weber APM, Yarmolinsky D, Zeni A, Cattivelli L (2025) Boosting photosynthesis opens new opportunities for agriculture sustainability and circular economy: the BEST-CROP research and innovation action. Plant J 121(3):e17264. https://doi.org/10.1111/tpj.17264
179. Demmig-Adams B, Koh S-C, Cohu CM, Muller O, Stewart JJ, Adams WW (2014) Non-photochemical fluorescence quenching in contrasting plant species and environments. In: Demmig-Adams B, Garab G, Adams III W, Govindjee (eds) Non-photochemical quenching and energy dissipation in plants, algae and cyanobacteria. Advances in photosynthesis and respiration, vol 40. Springer, Dordrecht, pp 531–552. https://doi.org/10.1007/978-94-017-9032-1_24
180. Liu M, Wang Y, Zhang H, Hao Y, Wu H, Shen H, Zhang P (2024) Mechanisms of photoprotection in overwintering evergreen conifers: sustained quenching of chlorophyll fluorescence. Plant Physiol Biochem 210:108638. https://doi.org/10.1016/j.plaphy.2024.108638
181. Demmig-Adams B, Adams III WW (2006) Photoprotection in an ecological context: the remarkable complexity of thermal energy dissipation. New Phytol 172(1):11–21. https://doi.org/10.1111/j.1469-8137.2006.01835.x
182. Bag P, Ivanov AG, Huner NP, Jansson S (2024) Photosynthetic advantages of conifers in the boreal forest. Trends Plant Sci. https://doi.org/10.1016/j.tplants.2024.10.018
183. Oluborode J, Chadzinikolau T, Formela-Luboińska M, Ye Z-P, Robakowski P (2025) Adaptive significance of age- and light-related variation in needle structure, photochemistry, and pigments in evergreen coniferous trees. Photosynth Res 163(1):3. https://doi.org/10.1007/s11120-024-01125-2
184. Croce R, Carmo-Silva E, Cho YB, Ermakova M, Harbinson J, Lawson T, McCormick AJ, Niyogi KK, Ort DR, Patel-Tupper D, Pesaresi P, Raines C, Weber APM, Zhu X-G (2024) Perspectives on improving photosynthesis to increase crop yield. Plant Cell 36(10):3944–3973. https://doi.org/10.1093/plcell/koae132
185. Keller B, Soto J, Steier A, Portilla-Benavides AE, Raatz B, Studer B, Walter A, Muller O, Urban MO (2024) Linking photosynthesis and yield reveals a strategy to improve light use efficiency in a climbing bean breeding population. J Exp Bot 75(3):901–916. https://doi.org/10.1093/jxb/erad416
186. Wang Y (2024) Improving photosynthetic efficiency in fluctuating light to enhance yield of C3 and C4 crops. Crop Environ 3(4):184–193. https://doi.org/10.1016/j.crope.2024.06.003
187. Slattery RA, Walker BJ, Weber APM, Ort DR (2018) The impacts of fluctuating light on crop performance. Plant Physiol 176(2):990–1003. https://doi.org/10.1104/pp.17.01234
188. Zhu X-G, Ort DR, Whitmarsh J, Long SP (2004) The slow reversibility of photosystem II thermal energy dissipation on transfer from high to low light may cause large losses in carbon gain by crop canopies: a theoretical analysis. J Exp Bot 55(400):1167–1175. https://doi.org/10.1093/jxb/erh141
189. Kaiser E, Walther D, Armbruster U (2020) Growth under fluctuating light reveals large trait variation in a panel of Arabidopsis accessions. Plants 9(3):316. https://doi.org/10.3390/plants9030316

190. Taylor SH, Long SP (2017) Slow induction of photosynthesis on shade to sun transitions in wheat may cost at least 21% of productivity. Philos Trans R Soc Lond B Biol Sci 372(1730): 20160543. https://doi.org/10.1098/rstb.2016.0543
191. De Souza AP, Burgess SJ, Doran L, Hansen J, Manukyan L, Maryn N, Gotarkar D, Leonelli L, Niyogi KK, Long SP (2022) Soybean photosynthesis and crop yield are improved by accelerating recovery from photoprotection. Science 377(6608):851–854. https://doi.org/10.1126/science.adc9831
192. Głowacka K, Kromdijk J, Kucera K, Xie J, Cavanagh AP, Leonelli L, Leakey ADB, Ort DR, Niyogi KK, Long SP (2018) Photosystem II subunit S overexpression increases the efficiency of water use in a field-grown crop. Nat Commun 9(1):868. https://doi.org/10.1038/s41467-018-03231-x
193. Hubbart S, Smillie IRA, Heatley M, Swarup R, Foo CC, Zhao L, Murchie EH (2018) Enhanced thylakoid photoprotection can increase yield and canopy radiation use efficiency in rice. Commun Biol 1(1):22. https://doi.org/10.1038/s42003-018-0026-6
194. Kromdijk J, Głowacka K, Leonelli L, Gabilly ST, Iwai M, Niyogi KK, Long SP (2016) Improving photosynthesis and crop productivity by accelerating recovery from photoprotection. Science 354(6314):857–861. https://doi.org/10.1126/science.aai8878
195. Garcia-Molina A, Leister D (2020) Accelerated relaxation of photoprotection impairs biomass accumulation in Arabidopsis. Nat Plants 6(1):9–12. https://doi.org/10.1038/s41477-019-0572-z
196. Lehretz GG, Schneider A, Leister D, Sonnewald U (2022) High non-photochemical quenching of VPZ transgenic potato plants limits CO2 assimilation under high light conditions and reduces tuber yield under fluctuating light. J Integr Plant Biol 64(9):1821–1832. https://doi.org/10.1111/jipb.13320
197. Küster L, Lücke R, Brabender C, Westhäuser S, Jahns P (2023) The amount of zeaxanthin epoxidase but not the amount of violaxanthin de-epoxidase is a critical determinant of zeaxanthin accumulation in Arabidopsis thaliana and Nicotiana tabacum. Plant Cell Physiol 64. https://doi.org/10.1093/pcp/pcad091
198. Faralli M, Lawson T (2020) Natural genetic variation in photosynthesis: an untapped resource to increase crop yield potential? Plant J 101(3):518–528. https://doi.org/10.1111/tpj.14568
199. Porcar-Castell A, Tyystjärvi E, Atherton J, van der Tol C, Flexas J, Pfündel EE, Moreno J, Frankenberg C, Berry JA (2014) Linking chlorophyll a fluorescence to photosynthesis for remote sensing applications: mechanisms and challenges. J Exp Bot 65(15):4065–4095. https://doi.org/10.1093/jxb/eru191
200. del Carmen García-Rodríguez L, Prado-Olivarez J, Guzmán-Cruz R, Rodríguez-Licea MA, Barranco-Gutiérrez AI, Perez-Pinal FJ, Espinosa-Calderon A (2022) Mathematical modeling to estimate photosynthesis: a state of the art. Appl Sci 12(11):5537. https://doi.org/10.3390/app12115537
201. Ort DR, Merchant SS, Alric J, Barkan A, Blankenship RE, Bock R, Croce R, Hanson MR, Hibberd JM, Long SP, Moore TA, Moroney J, Niyogi KK, Parry MAJ, Peralta-Yahya PP, Prince RC, Redding KE, Spalding MH, van Wijk KJ, Vermaas WFJ, von Caemmerer S, Weber APM, Yeates TO, Yuan JS, Zhu XG (2015) Redesigning photosynthesis to sustainably meet global food and bioenergy demand. Proc Natl Acad Sci 112(28):8529–8536. https://doi.org/10.1073/pnas.1424031112
202. Gamon JA, Somers B, Malenovský Z, Middleton EM, Rascher U, Schaepman ME (2019) Assessing vegetation function with imaging spectroscopy. Surv Geophys 40(3):489–513. https://doi.org/10.1007/s10712-019-09511-5
203. Kureel N, Sarup J, Matin S, Goswami S, Kureel K (2022) Modelling vegetation health and stress using hyperspectral remote sensing data. Model Earth Syst Environ 8(1):733–748. https://doi.org/10.1007/s40808-021-01113-8

Time-Resolved Fluorescence Spectroscopy in Plants

Volha Chukhutsina

Contents

Abstract Time-resolved fluorescence spectroscopy (TRFS) is a powerful, noninvasive technique for probing ultrafast energy transfer, structural organization, and charge separation processes within photosynthetic machinery in vivo. Furthermore, TRFS measurements on intact leaves provide valuable insights into the response of the photosynthetic machinery to light stress or other abiotic changes. TRFS enables the monitoring of the primary steps of photosynthesis on picosecond to nanosecond timescales. Despite its widespread application for isolated complexes and unicellular organisms, the use of TRFS on intact leaves presents unique technical challenges due to the complexity of the sample. These challenges include minimizing excitation-induced annihilation effects, ensuring well-defined photosynthetic states, maintaining leaf integrity, and accounting for spectroscopic artifacts caused by tissue heterogeneity, scattering, and light reabsorption. Therefore, proper experimental considerations and configurations are required to mitigate these factors and ensure

V. Chukhutsina (✉)
Biophysics of Photosynthesis, Department of Physics and Astronomy, Faculty of Science, Vrije Universiteit Amsterdam and LaserLaB Amsterdam, De Boelelaan, Amsterdam, The Netherlands
e-mail: v.u.chukhutsina@vu.nl

M. G. Lagorio (ed.), Comprehensive Series in Photochemical and Photobiological Sciences, Fluorescence of Living Plants 1, https://doi.org/10.1007/17361_2025_12
Published online: 17 July 2025

reliable and interpretable fluorescence decay measurements. TRFS thus remains an essential tool for advancing the understanding of photosynthetic performance in plants, plant stress responses, and strategies for improving crop productivity. This chapter reviews the principles, methodologies, and challenges associated with time-resolved fluorescence measurements in intact leaves, highlighting current advancements and knowledge gaps in the field.

Keywords Fluorescence, Leaf, Light reabsorption, Time-resolved spectroscopy

Abbreviations

A. thaliana	*Arabidopsis thaliana* (model C3 plant)
ADC	Analog-to-digital converter
CCD	Charge-coupled device
CFD	Constant fraction discriminator
CP24, CP26, CP29	Minor antenna proteins of PSII
DCMU	3-(3,4-Dichlorophenyl)-1,1-dimethylurea (PSII inhibitor)
$F(\lambda)$	Measured fluorescence spectrum at wavelength λ
F_0	Minimum fluorescence yield (PSII open state in dark-adapted state)
F_{685}	Fluorescence emission at 685 nm
F_{735}	Fluorescence emission at 735 nm
$F_{corr}(\lambda)$	Corrected fluorescence spectrum at wavelength λ
FLIM	Fluorescence lifetime imaging microscopy
F_m	Maximum fluorescence yield (PSII closed state in dark-adapted state)
F'_m	Maximal fluorescence yield (in light-adapted state)
F_{NPQ}	Fluorescence under non-photochemical quenching conditions
F_{ST1}, F_{ST2}	Fluorescence in state 1/state 2 transitions
F_v/F_m	Variable to maximal fluorescence ratio (indicator of PSII efficiency)
FWHM	Full width at half maximum
IRF	Instrument response function
ko	Knockout (genetic mutant)
LED	Light-emitting diode
LHCII	Light-harvesting complex II
M. giganteus	*Miscanthus giganteus* (C4 plant)
MCA	Multichannel analyzer
MCP	Microchannel plate
MCP-PMT	Microchannel plate photomultiplier tube
mpm	Motions per minute (lateral motion speed)
N. tabacum	*Nicotiana tabacum* (tobacco plant)

NPQ	Non-photochemical quenching
ns	Nanoseconds (10^{-9} s)
OPA	Optical parametric amplifier
OPO	Optical parametric oscillator
$P700^+$	Oxidized reaction center chlorophyll in PSI
PAM	Pulse-amplitude modulated fluorometry
pJ	Picojoules (10^{-12} J)
PMT	Photomultiplier tube
ps	Picoseconds (10^{-12} s)
PsbS	PSII subunit S (pH sensor protein)
PSI	Photosystem I
PSII	Photosystem II
qE	Energy-dependent quenching (component of NPQ)
qI	Photoinhibitory quenching (component of NPQ)
qT	State-transition quenching (component of NPQ)
RC	Reaction center
rpm	Rotations per minute
S–S	Singlet–singlet (annihilation)
S–T	Singlet–triplet (annihilation)
TAC	Time-to-amplitude converter
TCSPC	Time-correlated single photon counting
TPE	Two-photon excitation
TRFS	Time-resolved fluorescence spectroscopy
Zx	Zeaxanthin
μmol photons m^2 s^{-1}	Micromoles of photons per square meter per second (light intensity unit)

1 Introduction

Time-resolved fluorescence measurements are widely used to study the primary steps of photosynthesis and the organization of photosynthetic machinery in vivo in unicellular organisms [1–4]. For light energy to be stored via photosynthesis in plants, it must be absorbed by one of the pigments (carotenoids or chlorophylls) within the thylakoid membrane. The timescale of this sequence of events ranges from a few femtoseconds (required to excite the pigment from the ground to the excited state) to picoseconds (for energy transfer to the reaction centers (RCs)) and up to nanoseconds (to initiate charge separation in the RCs). Therefore, time-resolved spectroscopy is the ideal tool to assess such ultrafast photosynthetic processes.

Two common time-resolved fluorescence spectroscopy (TRFS) techniques for performing bulk experiments in vivo are streak camera [5] and time-correlated single-photon counting (TCSPC) [6], with TCSPC combined with microscopy approaches resulting in fluorescence lifetime imaging (FLIM) [7]. Fluorescence, emitted by chlorophyll a in oxygenic photosynthetic organisms, serves as a valuable, noninvasive probe of photosynthetic performance. By tracking fluorescence decay over picosecond-to-nanosecond timescales, researchers can resolve energy transfer from light-harvesting antenna complexes to RCs. Furthermore, TRFS provides insights into the distinct decay kinetics of photosystem II (PSII) and photosystem I (PSI), which differ due to variations in pigment composition and make it possible to resolve energy transfer pathways and charge separation mechanisms. This technique is crucial for assessing photosystem efficiency, electron transport dynamics, and regulatory processes such as non-photochemical quenching (NPQ) and state transitions. TRFS, therefore, contributes to a deeper understanding of photosynthetic mechanisms and adaptive responses of plants [8–10], algae [11, 12], and cyanobacteria [13] to changing environmental conditions.

One of the key TRFS advantages for plant research is its ability to study the processes in intact leaves under natural conditions, without the need for invasive or destructive methods. This makes it a valuable tool for examining how plants respond to environmental factors such as light intensity, temperature, drought, and nutrient availability [8, 9, 14]. Furthermore, the technique provides crucial insights into the PSI and PSII overall efficiency and performance, which is essential for improving crop performance, understanding plant adaptation strategies, and advancing sustainable agriculture.

Various laboratories have previously demonstrated the power of time-resolved fluorescence spectroscopy to reveal ultrafast photosynthetic dynamics in thylakoids [15], isolated photosynthetic complexes [12, 16, 17], and photosynthetic unicellular organisms such as algae [4, 11, 18] and cyanobacteria [13, 19]. However, time-resolved measurements on intact leaves are not yet well established, and a solid approach to perform such measurements has been developed recently [10, 20, 21]. This chapter reviews the current state of the art, methodologies, and knowledge gaps in this field.

2 Principles and Technical Challenges

2.1 Time-Resolved Spectroscopy on Intact Leaves: Considerations

Due to the high complexity of the samples, time-resolved fluorescence measurements on intact leaves require several considerations:

2.1.1 Annihilation-Free Conditions

Under high-light excitation densities, multiple excited states may arise within a single photosynthetic complex, leading to singlet-singlet or singlet-triplet annihilation in chlorophylls. These annihilation processes shorten the measured fluorescence decay traces [22]. Such dynamics can be well described using kinetic or statistical approaches in isolated antenna systems [23–25]. The annihilation kinetics depend not only on the excitation intensity but also on the diffusion radius, the number of pigments involved, and their connectivity [26]. As the presence of annihilation affects the fluorescence kinetics, it introduces unnecessary complexity into in vivo measurements. Accounting for these effects in data analysis is challenging [22, 24]. For this reason, annihilation should be avoided in TRFS experiments whenever possible.

2.1.2 Photosystem II Kinetics and Measuring Conditions

Charge separation in open PSII occurs 3–6 times faster than in closed PSII [17, 27], leading to significantly slower fluorescence kinetics in the closed state compared to the open state. Therefore, to interpret the results unambiguously, it is crucial to measure in a well-defined state of PSII: F_0, F_m, and F'_m.

2.1.3 Leaf Fragility and Environmental Stress

Intact leaves are delicate and susceptible to various abiotic stress factors such as changes in temperature, humidity, and mechanical stresses [28]. Special care should be taken with the leaf samples to prevent any changes during the experiment.

2.1.4 Spectroscopic Considerations

Intact leaves contain a variety of chromophores – including chlorophylls, carotenoids, phenolic compounds, flavonoids, and lignin – that are heterogeneously distributed throughout the sample depth. Combined with the complexity of leaf morphology, this distribution can introduce artifacts in measurements. The interaction between tissue absorbers and scatterers renders the tissue optically turbid. Previous studies have shown that fluorescence from biologically turbid tissues (with high number of absorbers and scatterers) could be distorted [29, 30].

Fluorescence spectra of intact leaves feature two main bands originated from chlorophyll a: one band in the red region (around 685 nm, F_{685}) and another in the near-infrared region (around 735 nm, F_{735}, Fig. 1). The 685 nm band is more susceptible to reabsorption due to greater overlap between absorption and emission spectra of chlorophyll a (Fig. 1). While reabsorption affects steady-state fluorescence spectra, it does not influence fluorescence kinetics [31, 32]. In heterogeneous, highly scattering samples such as animal tissues, minerals, or polymer blends, reabsorption can combine with other optical effects like reemission and time-of-flight dispersion, which can alter fluorescence decay traces [31]. Reemission has been shown to slow

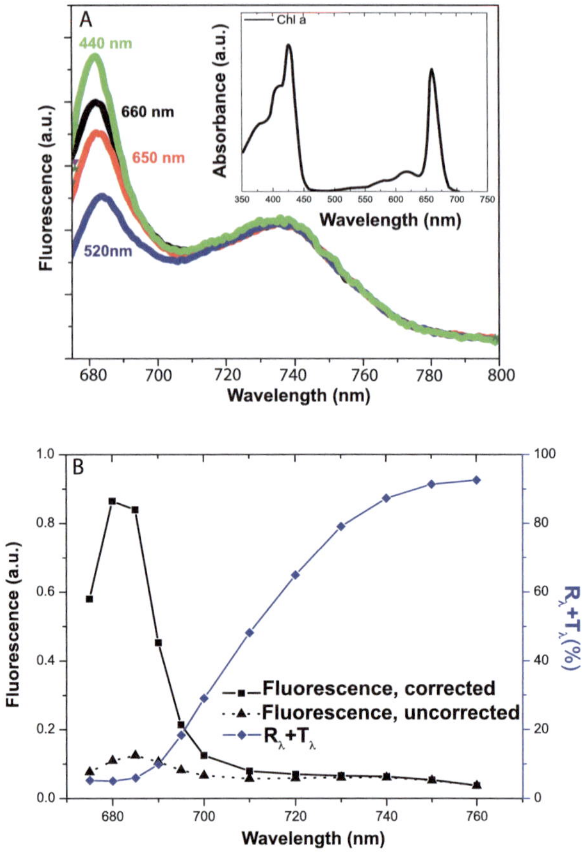

Fig. 1 (**a**) Steady-state fluorescence spectra of intact leaves at different excitation wavelengths, showing the dependence of the reabsorption on the sample transmission properties (T) at the excitation wavelength, as described by Eq. (1) (Inset: Absorbance spectrum of chlorophyll a). (**b**) Steady-state fluorescence at 660 nm excitation of *Arabidopsis thaliana* leaves before (black dotted line) and after (black solid line) reabsorption correction. $R_\lambda + T_\lambda$ measured on intact *A. thaliana* leaves is presented by blue solid line

fluorescence kinetics, particularly in concentrated free chlorophyll solutions [31, 33, 34]. However, this effect can be minimized with thin layers of photosynthetic tissue [31] or front-face detection [33]. Since only 1–2% of absorbed light is emitted as fluorescence in intact leaves [35], the impact of reemission was experimentally demonstrated to be negligible: TRFS measurements on intact leaves, when performed in front-face detection mode, demonstrate that fluorescence kinetics remain undistorted, indicating that the fluorescence is collected from the thin top layer of the leaf tissue [21].

2.1.5 Light Reabsorption Effect

Correcting the time-resolved spectra for reabsorption is recommended since it enhances the interpretation of the data [36]. Several models have been proposed to account for reabsorption of fluorescence spectra [29, 34]. Gitelson and coworkers [37] experimentally demonstrated that the true F_{685}/F_{735} ratio of steady-state fluorescence in intact leaves is closely related to the ratio of non-reabsorbed light at 685 and 735 nm. Later models [38–40] employed defined physical approaches to describe the interaction between light and matter, offering a more precise quantification of reabsorption. Both the Agati et al. [38] and Lagorio et al. [39] models were shown to be well suited for reliably accounting for reabsorption. Among these, the model by Agati and colleagues is considered preferable for measurements on unstacked leaves. Based on it [38], fluorescence spectra ($F(\lambda)$) can be corrected for light reabsorption ($F_{\text{corr}}(\lambda)$) using the following equation:

$$F_{\text{corr}}(\lambda) = F(\lambda) * f,$$

where

$$f = \frac{\ln \frac{1}{\left(R_{\lambda_0} + T_{\lambda_0}\right)} + \ln \frac{1}{\left(R_{\lambda} + T_{\lambda}\right)}}{\ln \frac{1}{\left(R_{\lambda_0} + T_{\lambda_0}\right)}} * \frac{1 - R_{\lambda_0} - T_{\lambda_0}}{1 - (R_{\lambda_0} + T_{\lambda_0})(R_{\lambda} + T_{\lambda})} \tag{1}$$

In this equation, R and T stand for reflected and transmitted radiation, respectively, while λ_0 and λ are the excitation and detection wavelength, respectively (Fig. 1). From Eq. (1), it follows that the effect of reabsorption is greatest when the excitation wavelength corresponds to the region of the weakest absorption, where $1 - (R_{\lambda_0} + T_{\lambda_0})$ values are the lowest (Fig. 1, insert). Strong absorption at the excitation wavelength limits light penetration, causing fluorescence to originate near the surface, where it experiences minimal reabsorption. In contrast, when absorption is weaker, light penetrates deeper into the tissue, and the resulting fluorescence must travel a longer path to exit, increasing the likelihood of reabsorption. However, some degree of light reabsorption occurs at all excitation wavelengths and must be carefully considered when interpreting fluorescence spectral signatures.

2.2 *Historical Perspective*

Time-resolved fluorescence measurements on intact leaves were first reported in the 1980s and 1990s by multiple research groups [41–43]. A significant advancement came from Moya and colleagues, who made the first attempts to conduct time-resolved measurements on intact leaves in photosynthetically defined states with sufficient time resolution (10–15 ps) [42, 43]. The leaves were measured in the F_s state (steady-state fluorescence, a mixed state with contributions from both open and closed PSII) and the F_m state (maximum fluorescence, with reduced PSII after DCMU infiltration). The fluorescence kinetics observed were multi-exponential, with lifetimes ranging from 25 to 850 ps for the F_s state and 30 ps to 2.4 ns for the F_m state, where the slowest components (>100 ps) were primarily affected by PSII closure. These data provided the first indication that spectral separation of PSII and PSI kinetics could be achieved in time-resolved fluorescence measurements on intact leaves. However, because the measurements were performed at the same spot, the F_0 state (the state with minimal fluorescence intensity, corresponding to fully oxidized PSII) could not be achieved. Additionally, the data may have contained secondary effects and artifacts induced by the laser, such as singlet-singlet and singlet-triplet annihilation, which can occur even at moderate laser powers [24].

In the 2010s, a new method for time-resolved spectroscopic measurements on intact leaves was introduced. This method utilizes a special rotational cuvette, connected to a moving motor, allowing for fast rotation and side movement of the measured leaves [14, 44]. The first experiments that employed it reported the ability to achieve defined photosynthetic states and perform detailed analysis of the measured data. Later, the proper methodology of such measurements has been established, and possible spectroscopic artifacts in time-resolved measurements on intact leaves were excluded [21].

2.3 Time-Resolved Fluorescence Versus Standard Physiology Characterization

A common approach for assessing various photosynthetic parameters in intact leaves involves measuring fluorescence yields in different states using pulse-amplitude modulated (PAM) fluorometry [45] or the Joliot-type spectrometer [46] (JTS-10). Recently, an extension of PAM called ChloroSpec [47] has been introduced, enabling measurements with spectral resolution [47]. All these techniques generate signals proportional to fluorescence yield, providing a high signal-to-noise ratio. To estimate the maximum quantum efficiency of PSII photochemistry, referred to as "the PSII maximum efficiency," the minimum and maximum chlorophyll (Chl) fluorescence levels of dark-adapted leaves are measured. Specifically, the photosynthetic sample is measured with either open PSII (F_0, oxidized PSII, corresponding to the dark-adapted state) or fully closed PSII (F_m, reduced PSII, typically achieved by a strong saturation pulse of at least 2,000 μE and 40–200 ms duration) [45, 47]. These correspond to minimal (F_0) and maximal (F_m) fluorescence yields, respectively. Maximum potential PSII efficiency, F_v/F_m, is then calculated as $(F_m - F_0)/F_m$.

Many other photosynthetic parameters are estimated under constant light illumination. For example, PSII operating efficiency is estimated as $\left(F'_m - F'\right)/F'_m$, where F'_m and F' are the maximal and minimal fluorescence levels under constant illumination. Various studies have demonstrated a strong linear relationship between PSII operating efficiency and CO_2 assimilation (e.g., [48–50]). Several coefficients are widely used by the photosynthetic community to account for different mechanisms initiating quenching of fluorescence levels upon illumination: qE (photoprotective deactivation of absorbed energy through heat via non-photochemical quenching, NPQ), qT (state-transitional changes), and qI (PSII photoinhibition).

All the interpretations of photosynthetic parameters mentioned above are greatly facilitated by the assumption that F_0, F', F_m, and F'_m values at room temperature originate mainly from PSII. However, there is a contribution from PSI fluorescence, which can be of considerable practical relevance. For example, in *Arabidopsis thaliana*, the PSI contribution is around 3% in the F_m state and 20% in the F'_m state [8]. Pfündel analyzed the fluorescence characteristics of 12 different species and estimated that the PSI contribution to total F_0 fluorescence at wavelengths greater than 700 nm is approximately 30% in C3 plants and 50% in C4 plants [51].

Additionally, under certain conditions, PSI kinetics may be significantly altered. For instance, direct energy transfer from PSII to PSI (spillover) has been proposed as an NPQ mechanism occurring in plants, algae, and cyanobacteria [3, 9, 52]. If present, spillover would completely alter both PSI and PSII kinetics [9], meaning that the calculated photosynthetic coefficients such as qI, qE, or qT can no longer be considered as originating solely from PSII.

Moreover, estimating the photosynthetic coefficients from PAM, JTS-10 or ChloroSpec would not provide insights into the actual mechanisms underlying changes in light harvesting, charge separation, and their regulation in response to environmental changes. For example, three possible mechanisms of NPQ, a major response of plants to rapid light increases, are suggested in the literature: Zx-related quenching, presumably localized in the PSII core [44]; disconnection of aggregated and quenched antenna [53]; and spillover [54]. These mechanisms, if present, would all result in a decrease of F'_m upon illumination. However, only time-resolved spectroscopy measurements can distinguish between these three NPQ mechanisms [9, 10].

Therefore, only TRFS measurements on intact leaves can provide insights into the primary steps of photosynthesis in vivo, uncovering quenching sites and revealing potential energy reorganization of the photosynthetic machinery in response to environmental changes (e.g., light adaptation, NPQ, state transitions, temperature fluctuations). This **chapter** aims to describe the basic principles, considerations, and recent developments in time-resolved fluorescence spectroscopy on intact leaves.

3 Time-Resolved Fluorescence Methods

3.1 Bulk Time-Resolved Fluorescence Methods

TRFS measurements are widely applied to study the primary steps of photosynthesis and the organization of photosynthetic machinery in vivo in unicellular organisms [1–4]: Time-resolved fluorescence spectroscopy allows for the observation of ultrafast dynamic events in photosynthetic systems in vivo on a ps-ns timescale by directly measuring fluorescence lifetimes. The fluorescence lifetime reflects the time a chlorophyll molecule spends in the excited state before returning to the ground state by emitting a photon. Fluorescence decay traces obtained from whole leaves represent an average of many lifetimes from various chlorophylls in different photosynthetic subunits (antenna, PSs, RCs). As a result, fluorescence decay kinetics of PSII and PSI and the corresponding light-harvesting antenna can be resolved separately: their fluorescence decays at different rates due to the presence of different pigment pools in combination with different charge separation kinetics in their RCs. Also, the spectra of the pigment pools in the antenna of PSII and PSI differ. That is why TRFS combined with proper data analysis [55, 56] can be used to probe the state and the organization of the photosynthetic apparatus. The two common techniques used to perform TRFS in vivo on unicellular phototrophs (e.g., green algae, diatoms) are the streak camera and time-correlated single photon counting (TCSPC). Both of them have been applied to perform measurements on intact leaves. The methodology of these two techniques is described below.

3.1.1 Time-Correlated Single Photon Counting

TCSPC (Fig. 2a) measures the time delay between excitation pulses and the subsequent emission of a fluorescence photon. Single photons are detected over many excitation cycles, and the time-dependent fluorescence signal is reconstructed from a histogram of delay times. To ensure accurate photon detection, laser power in TCSPC is kept low, detecting only one photon per excitation pulse.

Most of the TCSPC setups are operated in “reverse” mode: This means that the photon detection triggers the TCSPC electronics to start the clock. The clock is stopped when the laser excitation pulse is measured. In this way, the time between photon detection and delayed excitation pulse is measured. After a short laser pulse excites the sample, the emission of a single fluorescence photon is detected by a microchannel plate-photomultiplier tube (MCP-PMT). A constant fraction discriminator (CFD) generates an electronic pulse that triggers a time-to-amplitude converter (TAC), with the excitation pulse acting as a stop signal after a fixed delay. The signal-to-noise ratio can be further improved by gating the TAC with the output of a differential discriminator, which selects only those pulses whose amplitudes fall within a predefined window between specified minimum and maximum values. The TAC develops a voltage that increases linearly over time, which is then converted into a numerical value by an analog-to-digital converter (ADC) and sorted into channels by a multichannel analyzer (MCA). The channel time spacing is typically set in the range of 2.0–5.0 ps.

To avoid distortion due to multiple photon detections per excitation pulse (pileup distortion), the fluorescence signal must be kept low, typically at less than 1 photon per 100 excitation pulses. This is achieved by reducing the excitation pulse energy to the pJ level. Therefore, for example, if a 4 MHz laser source is used, the photocounting rate should be limited to about 40,000 photons s^{-1}.

The most accurate TCSPC setups use mode-locked Ti:sapphire lasers, typically combined with an optical parametric oscillator (OPO) or optical parametric amplifier (OPA) to provide a sub-ps pulse of the variable wavelengths in the visible range. A

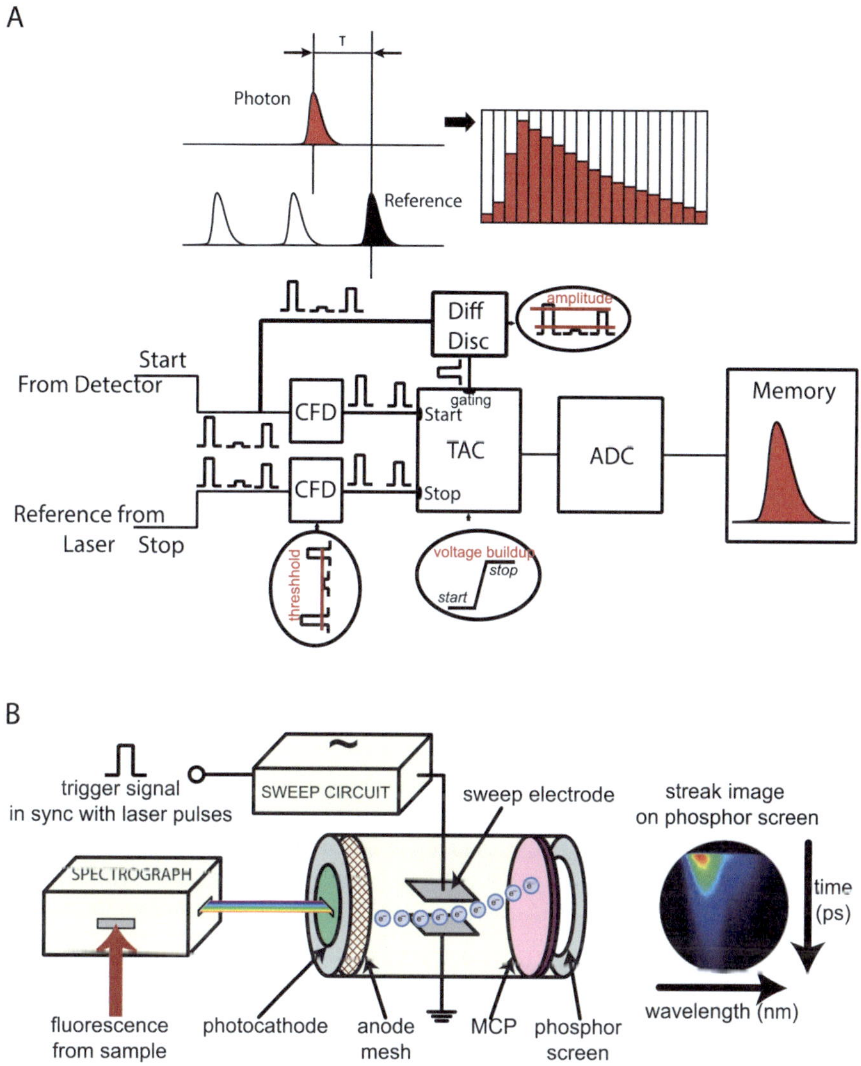

Fig. 2 Schematics of two major time-resolved fluorescence approaches used to study intact leaves. (**a**) Time-correlated single photon counting scheme. (**b**) Streak camera scheme. See Sect. 3 for more information

pulse picker reduces the repetition rate to 0.5–4 MHz range. For decay trace measurements at various wavelengths, interference filters or monochromators are employed. There are also commercially available single-box TCSPC systems that combine all the electronics described above and offer high resolution, typically below 50 ps [57].

3.1.2 Streak Camera Setup

Another widely used time-resolved fluorescence technique involves detection with a synchroscan streak camera system combined with a spectrograph (Fig. 2b) [5, 58]. This system disperses photoelectrons across an imaging screen in both wavelength and time, allowing the monitoring of the entire spectral evolution on a picosecond-to-nanosecond timescale, rather than measuring single fluorescence decay curves at individual wavelengths [59]. Additionally, a streak camera offers superior time resolution compared to TCSPC. The full-width-at-half-maximum (FWHM) of the instrument response function (IRF) for the streak camera can reach below 1 ps [5], while the TCSPC setup with MCP PMT detection typically has an FWHM of at least 10–20 ps [9].

Figure 2b illustrates the operating principle of a common streak camera setup used for photosynthesis research applications [11, 60, 61]. Fluorescence photons emitted by laser excitation are focused by an objective onto the input slit of a spectrograph, where they are dispersed by a grating in the horizontal direction according to their wavelength. The photons then strike the photocathode, causing the emission of electrons. These photoelectrons first pass through an accelerating mesh, after which they travel between a pair of sweep electrodes. A high sweeping voltage is applied to the electrodes, and the sweep frequency is synchronized with the frequency of the excitation light pulses. During the high-speed sweep, electrons arriving at slightly different times are deflected at varying angles in the vertical direction, entering the micro-channel plate (MCP) at different positions. As the photoelectrons strike the MCP, they are multiplied several thousand times, after which they reach a phosphor screen, where they are again converted into light. The photoelectrons generated at different times experience different deflection fields, causing them to strike the phosphor screen at distinct vertical positions. The emission from the phosphor screen is then focused onto a CCD detector, and the data is stored for further processing. A mode-locked Ti:sapphire laser is usually used to generate short (sub-ps) light pulses. A small fraction of the laser output should be used for synchronization with the streak camera's sweep field.

The main advantage of streak camera measurements is the simultaneous detection of both photon emission time and wavelength [5]. In contrast, TCSPC provides an excellent signal-to-noise ratio and very high sensitivity, which is crucial for measuring in low fluorescent states, like F_0 or F'_{m}. Given the highly complex nature of time-resolved fluorescence kinetics in intact leaves [9, 36], obtaining fluorescence decay traces with a high signal-to-noise ratio is essential for extracting meaningful information from the resolved fluorescence data. Furthermore, TCSPC detection window is limited only by the repetition rate of the laser. When using a repetition rate below 5 MHz, TCSPC enables the detection of fluorescence traces over the entire fluorescence decay range (hundreds of nanoseconds). In contrast, streak cameras typically have a more restricted time window of 1.5–2 ns, limiting the ability to capture longer-lived fluorescence components. Therefore, TCSPC usually provides excellent time separation for PSI and PSII kinetics. For these reasons, TCSPC is often the method of choice for leaf measurements [9, 10, 21]. However, the streak camera, with its ability to combine spectral and time information simultaneously, could be the preferred method in certain cases [20].

3.1.3 Measuring on Intact Leaves Using TRFS Approach

To study the photosynthetic response of intact leaves, it is generally beneficial to measure fluorescence from the upper side of the leaf (facing the sun), where the primary photosynthetic tissue, the palisade mesophyll, is located. Due to high scattering and reabsorption, time-resolved measurements should be conducted in a front-face configuration. However, to achieve defined photosynthetic states, the sample volume must be constantly refreshed. In TRFS measurements on single-celled organisms or isolated complexes, this is typically achieved using a flow cell [18]. For leaves, this can be accomplished with a custom-made rotational-translational cuvette (Fig. 3). The first design was developed in the group of Prof. A. Holzwarth, where this cuvette consisted of two glass plates with a rubber ring between them, tightly sealed with two metal plates [10]. Detached plant leaves are placed between the glass plates and mounted in the cuvette (diameter = 10 cm, thickness = 0.5–4 mm). The cuvette is rotatable, with typical rotation speeds ranging

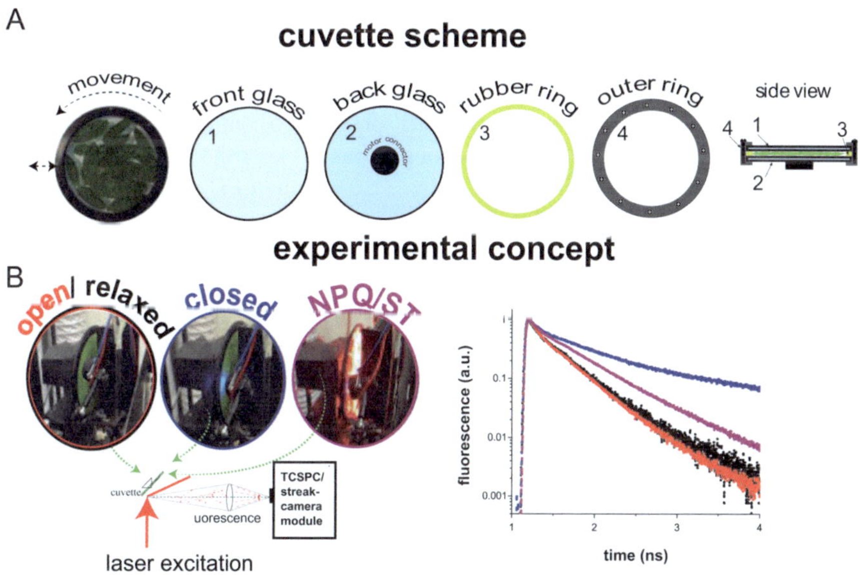

Fig. 3 (**a**) The rotation cuvette design for measurements on intact leaves. Dimensions of the cuvette: diameter 10 cm, thickness 0.5–4 mm. (**b**) The concept of TRFS measurement on intact leaves in different physiological states: open (F_0) or closed (F_m) photosystem II (corresponding to minimum and maximum fluorescence levels of PSII, which occur when the electron transport carriers from PSII are in oxidized or reduced state, respectively), in the presence of NPQ or state transitions (STs) (achieved by continuous illumination with dedicated LEDs). The relaxed state refers to the condition in which leaves recover in darkness after specific light treatment (NPQ or ST)

from 600 to 1,200 rpm, depending on the plant species [36]. It can also oscillate laterally at speeds of 60–80 mpm. This motion and rotation of the cuvette ensure that the average laser intensity per sample volume remains very low, thereby excluding secondary effects induced by the laser.

Additionally, it is important to ensure that the cuvette can be hermetically sealed: due to centrifugal force, liquid may leak out of the sample, potentially leading to dehydration and stress in the leaf tissue. To prevent this, the leaves should be immersed in a sucrose solution during measurements. By combining low excitation powers with rotation and lateral motion, TRFS measurements are achievable in any physiologically relevant state of the photosynthetic apparatus.

Achieving Well-Defined Photosynthetic States

The physiologically relevant states of the photosynthetic machinery are typically the following (Fig. 3b): open (F_0), closed (F_m), quenched (F_{NPQ}), and state transitions (F_{ST1}, F_{ST2}):

1. **Open state (F_0)**: F_0 is measured in complete darkness after some dark adaptation. The excitation energy of the laser pulse for open-state detection is typically in the pJ range. It is essential to perform preliminary checks using different laser pulse energies, repetition rates, and rotation speeds to ensure that the PSII RCs are open and there is no leaf damage due to rotation.
2. **Closed state (F_m)**: To measure leaves with closed PSII reaction centers, dark-adapted leaves are preincubated in DCMU solution. To confirm full closure of the PSII RCs, a weak blue LED (~50 μmol photons m^2 s^{-1}) should be focused in a small spot above the excitation site (Fig. 3). The blue LED intensity should be adjusted to the minimum value that still allows the maximum fluorescence yield but does not induce NPQ or other effects.
3. **Quenched state (F_{NPQ})**: Light adaptation to reach non-photochemical quenching (NPQ) is typically done with a high-intensity red LED panel emitting 800 μmol photons m^2 s^{-1}. Measurements should start only after the quenching level has stabilized (10–30 min, depending on illumination intensity and plant species). To fully close PSII RCs under quenched conditions, a high-intensity blue LED is focused on the excitation site. For F_{NPQ} state measurements with open RCs, Farooq and colleagues [20] used a protocol in which the time-resolved fluorescence signal is measured with the streak camera in the presence of stable NPQ for 10 s while actinic light is blocked. After recording time-resolved fluorescence signal for 10 s, the shutter for actinic light illumination opens and the shutters for the laser and detector get blocked. In this way, the sample gets re-illuminated with actinic light for 1 min to stabilize NPQ [20]. After NPQ stabilization, time-resolved fluorescence signal is remeasured again for another 10 s. This procedure is repeated until an acceptable signal-to-noise ratio is achieved.
4. **State transitions**: To achieve state I, the leaves should be illuminated for 10–20 min with low-intensity red light (PSII light). To achieve state II, low far-red light should be provided [62, 63].

Power and Stability Study

To accurately measure the photosynthetic machinery in a defined state, several checks must be performed. When using a cuvette, it is crucial to assess sample stability, as leaf tissue is susceptible to mechanical stress [64]. Excessive rotation speed or prolonged experiments can cause leaf damage, which may affect time-fluorescence kinetics. An additional control could involve measuring the F_v/F_m value [45], a commonly used indicator of plant fitness, both before and after the measurement to ensure that it remains unchanged. Alternatively, a fluorescence decay trace at a characteristic wavelength should be measured before and at the end of the measurement (Fig. 4). A power dependency study is also essential to confirm that measurements are conducted in the F_0 state, eliminating any laser-

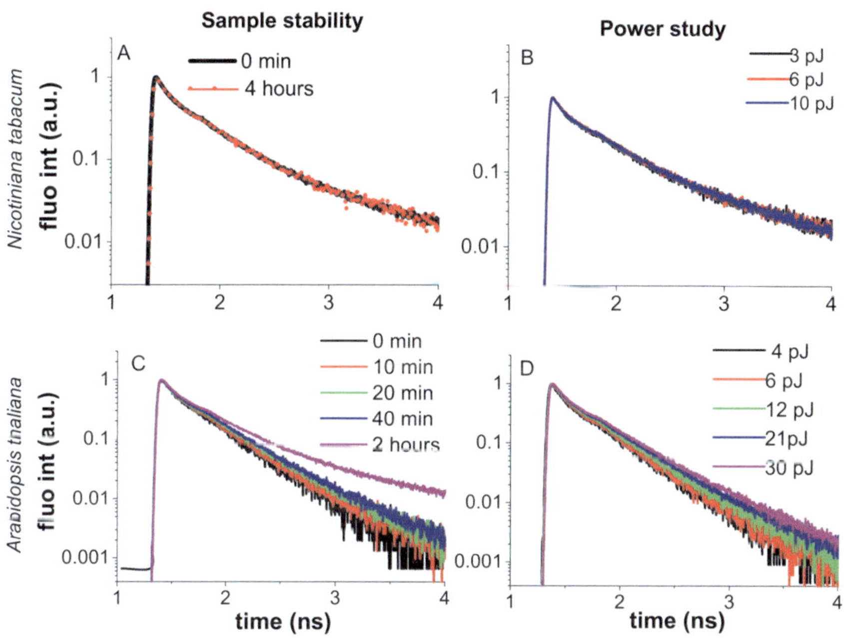

Fig. 4 Sample stability and excitation energy checks should always be performed prior to conducting time-resolved fluorescence spectroscopy (TRFS) on intact leaves (modified from [21]). These checks should cover the characteristic wavelength range of 680–686 nm, which corresponds closely to the emission maxima of PSII and free chlorophyll. (**a**, **b**) Measurements on *Nicotiana tabacum* leaves in the F_0 state, at a rotation speed of 1,000 rpm: (**a**) Sample stability was confirmed by comparing fluorescence decay traces over a 4-h period, which showed no detectable changes in fluorescence kinetics. (**b**) An energy titration study demonstrated that fluorescence kinetics remain unaffected within the excitation energy range of 3–10 pJ. (**c**, **d**) *Arabidopsis thaliana* leaves require different experimental conditions to reach the F_0 state compared to *N. tabacum*: (**c**) At a rotation speed of 1,000 rpm, irreversible damage to the leaves was observed after only 40 min. Experimental testing indicated that a maximum speed of 600 rpm should be used to prevent damage during 4-h measurements. (**d**) Energy titration showed that an excitation energy of no more than 6 pJ is sufficient to achieve the F_0 state when rotating at 600 rpm

related artifacts. The detection wavelength used for such controls should be selected near the emission peak of free chlorophyll (to detect leaf damage) or PSII emission maxima (to confirm PSII closure).

For example, experiments conducted on *A. thaliana* at the same rotational speed (1,000 rpm) used for *Nicotiana tabacum* showed significant damage to the photosynthetic machinery of *A. thaliana* leaves, with fluorescence decay traces becoming longer-lived after just 40 min of measurement. Consistent with this, the F_v/F_m ratio of *A. thaliana* leaves dropped from 0.8 to 0.4 after 2 h of rotation at 1,000 rpm. Thus, a slower rotation speed is necessary for obtaining accurate measurements in *A. thaliana*. After testing various speeds, it was found that 600 rpm does not cause damage to *A. thaliana* leaves within the experimental duration (4–6 h) [21]. However, the slower rotation speed for *A. thaliana*, compared to *N. tabacum*, results in 1.7 times higher average power per sample volume. Therefore, to achieve the F_0 state in *A. thaliana*, the measuring power should be reduced compared to *N. tabacum* at 1,000 rpm rotation, and a power dependency study is necessary.

3.1.4 Fluorescence Lifetime Imaging

Fluorescence lifetime imaging (FLIM) is an advanced imaging technique that combines time-correlated single photon counting (TCSPC) with microscopy. In many cases, photosynthetic samples are spatially inhomogeneous, with varying fluorescence kinetics across different regions [65, 66]. In such cases, bulk experiments provide only limited information, making it more advantageous to combine TCSPC with microscopy. This integration results in a setup known as fluorescence lifetime imaging microscopy (FLIM). In FLIM measurements, the excitation pulse is focused on a small-volume element of the sample (referred to as a voxel in three dimensions and a pixel in two dimensions), and the resulting fluorescence is detected by a TCSPC detection module. The excitation beam is scanned across the sample in the *xy*-plane, and the fluorescence decay curve is recorded for each individual pixel. FLIM setups used in photosynthesis research typically employ either one-photon [66] or two-photon excitation (TPE) [65, 67, 68] to excite the sample. In 1931, Maria Göppert-Mayer theoretically demonstrated that two quasi-simultaneous photons could excite a molecule [69]. For TPE to occur, two photons must interact with a pigment within approximately 1 fs (10^{-15} s). Since the probability of TPE occurring under typical light conditions is low, special precautions are required to ensure that the photon density in both time and space is sufficiently high. This can be achieved by using ultrashort, high-energy light pulses in a tightly focused laser beam. While TPE requires more effort to implement, it offers several advantages. The penetration depth of near-infrared light used in TPE is significantly greater than that of visible light used in one-photon excitation microscopy [70]. Furthermore, excitation occurs only in the small spot with the highest photon density, eliminating the need for a confocal pinhole. The laser output is coupled into the microscope. The excitation light is focused into the sample using an objective lens. Fluorescence is detected by the same objective and separated from the excitation light by a dichroic mirror.

Single photons are then detected by a PMT detector, similar to the one used in TCSPC experiments.

3.2 *Time-Resolved Spectroscopic Insights into Photosynthetic Organization and Regulation In Vivo*

Although TRFS on intact leaves is a relatively new approach, it has significantly advanced our understanding of photosynthesis organization and activity in vivo. The key research findings are discussed below.

3.2.1 Bulk TRFS Experiments

Non-photochemical quenching (NPQ) is a key photoprotective mechanism in plants, regulated by PsbS [71], a pH-sensing protein, and the xanthophyll carotenoid zeaxanthin (Zx) [72].

Early in vivo and in vitro studies on NPQ suggested that both PsbS and zeaxanthin act only at the level of PSII, while PSI does not require additional photoprotection because P700+ serves as an effective quencher [73]. However, this view was challenged in 2014 by Ballottari et al., who studied PSI–LHCI complexes and reported that zeaxanthin-dependent quenching results in a 30% reduction in the functional antenna size of PSI [74]. To address this question in vivo, PSI kinetics were analyzed using the TRFS approach on intact leaves, revealing no effect of NPQ on PSI kinetics [8]. Further investigation of PSI organization and diversity in vivo using TRFS demonstrated that the antenna size in plants is up to 50% larger than previously estimated, based on photophysical and biochemical studies of isolated complexes [36]. These findings also indicated that, contrary to previous assumptions, there is natural biodiversity in PSI sizes – the number of additional LHCII complexes attached to PSI varies among plant species.

Using TRFS on intact *Arabidopsis thaliana* leaves, Holzwarth and colleagues examined the effect of NPQ on PSII in the PsbS-deficient *npq4* mutant (*psbs-1.3*), the PsbS-overexpressing mutant L17, and the *npq1* mutant (which lacks violaxanthin de-epoxidase and is unable to accumulate Zx in high light) [14]. This study identified two distinct NPQ quenching sites in vivo, each with different dependencies on PsbS and Zx. The first site (Q1) was proposed to be located in the functionally detached major light-harvesting complex of PSII and to depend solely on PsbS. The second site (Q2) was concluded to reside in the minor antennae of PSII and to require Zx for quenching.

This two-site NPQ model was further refined through TRFS measurements on additional mutants lacking PSII minor antennae complexes, such as the knockout (ko) mutants *koCP26* and *koCP29* and the double mutant *koCP24/CP26* [44]. These measurements confirmed the previously proposed NPQ model and demonstrated that the second quenching site (Q2), located in the PSII-attached antenna, is highly dependent on Zx and substantially relies on the CP29 complex for effective Q2 quenching. In the absence of CP29, Q2 quenching is significantly reduced [44].

Farooq and colleagues, studying fluorescence kinetics during NPQ with open and closed PSII reaction centers (RCs), proposed that the NPQ rate depends on the state of PSII reaction centers (RCs). Their findings suggest that NPQ is highly efficient when PSII RCs are closed and less efficient when PSII RCs are open. These results support the existence of at least two quenching mechanisms proposed earlier. One of these, based on its spectroscopic signature, resembles Q1. It relaxes very quickly (within 10 s) and is absent when PSII is open. The far-red band associated with this fast-relaxing NPQ state appears only when reaction centers are closed, disappearing within 1 min after turning off the actinic light or upon RC opening. The proposed PsbS-related origin of this fast-relaxing NPQ state, as suggested by the Holzwarth group [14, 53], still requires further confirmation. Farooq and colleagues argue that such rapid regulation (on the scale of seconds) and its dependence on the RC state are unlikely to be mediated by functionally disconnected LHCII. Instead, they propose that the quencher must be located in close proximity to the reaction center for effective quenching. However, the underlying molecular mechanism of this dynamic NPQ process remains to be elucidated.

TRFS is also a valuable technique for studying the effects of temperature stress on the photosynthetic machinery. Bag and colleagues [9] investigated cold stress in Scots pine and demonstrated that cold stress triggers efficient quenching via a spillover mechanism. Their study provided insight into the molecular mechanism of overwintering in evergreens – at least in Scots pine – where P700+ in PSI quenches PSII excitation after PSII comes into close contact with PSI, allowing direct energy transfer.

Thus, TRFS on intact leaves is an essential tool for studying photosynthetic organization and photoregulation in vivo.

3.2.2 Time-Resolved Fluorescence Microscopy

Two-photon fluorescence lifetime imaging microscopy (FLIM) was employed to analyze the distribution and properties of PSI and PSII in palisade and spongy chloroplasts in intact leaves of *Arabidopsis thaliana* (C3 plant) and *Miscanthus giganteus* (C4 plant) [75]. The FLIM approach was applied to achieve time-resolved-based separation of PSI and PSII. The results revealed a larger PSII antenna size on the abaxial side of *A. thaliana* leaves, likely due to shading by palisade cells in the spongy mesophyll, while PSI chlorophyll content was slightly higher on the adaxial side. In *M. giganteus*, mesophyll and bundle sheath cells exhibited distinct PSI/PSII ratios, with the PSI-to-PSII ratio in bundle sheath cells being at least 2.5 times higher than in mesophyll cells. All in all, these findings indicate that FLIM is an effective tool for studying the PSI/PSII ratio and PSII antenna size in different regions of plant leaves without requiring chloroplast isolation. This approach was further validated in 2017 by Wientjes and colleagues [76], who used FLIM measurements to quantify the PSI/PSII ratio in chloroplasts within their natural leaf environment. In intact leaves of *A. thaliana* plants grown under white light, the PSI/PSII chlorophyll ratio was 0.99 ± 0.09 on the adaxial side and 0.83 ± 0.05 on the abaxial side. After acclimation to far-red light, the PSI/PSII ratio decreased by more than threefold to compensate for PSII's ineffective absorption of far-red light, illustrating how plants optimize light harvesting according to their light conditions. This indicates that FLIM can be used for studying photoregulation in intact leaves. Verhoven and co-workers used confocal fluorescence lifetime imaging to quantify state transitions in *A. thaliana* chloroplasts in situ with sub-micrometer resolution [77]. They demonstrated that for all wavelengths, the PSI/(PSI + PSII) excitation ratio was higher in state 2 than in state 1. Imaging the local PSI/(PSI + PSII) excitation ratio in individual chloroplasts confirmed that LHCII migrates between the grana and stroma lamellae during state transitions. Fluorescence intensity images showed that thylakoid macro-organization remains largely unchanged during state transitions. All in all, single chloroplast imaging in intact leaves combined with the FLIM approach provides valuable insight into how plants adjust their photosynthetic machinery to dynamic light conditions [65, 77].

4 Open Questions and Directions of Future Research

Despite significant progress in the field over the past decade, many fundamental questions about photosynthesis remain unanswered, particularly those that must be addressed using the TRFS approach at the whole-leaf level. One of the main challenges lies in the complexity and fragility of leaf tissues, which are highly sensitive to measurement conditions. Consequently, developing reliable measurement protocols and models for data interpretation is often difficult and requires advanced methodologies. While the photochemical and photoregulatory processes governing photosynthesis have been extensively studied, several aspects remain unresolved.

For instance, although various models have been proposed to explain the mechanisms of photoregulation – such as state transitions and non-photochemical quenching (NPQ) – some aspects of these essential processes remain unclear, highlighting the need for TRFS on intact leaves to resolve them unambiguously. Specifically, the dynamics of NPQ induction and deactivation have never been examined using the TRFS approach in situ, necessitating further investigation. Initial TRFS studies on NPQ dynamics on intact leaves have yielded intriguing results, suggesting highly dynamic regulation of photochemistry and NPQ levels in response to excitation pressure [20].

Despite substantial progress in understanding the reorganization of PSI and PSII complexes during state transitions [78] – achieved through various biochemical and biophysical approaches [79] – most TRFS studies addressing the structural and functional arrangements of newly formed state transition complexes have been conducted on isolated photosynthetic subunits [80]. Although FLIM can capture changes in PSI/(PSI + PSII) ratios and PSII antenna size in situ [76, 77], comprehensive bulk TRFS studies during state transitions are still needed. These are essential to clarify the average number of antenna complexes involved, changes in energy transfer pathways between the two photosystems, and alterations in the efficiencies of PSI and PSII.

Another critical area of research involves the effects of temperature stress, as well as abiotic and biotic stressors, on photosynthesis. Recent TRFS studies have shown that cold-induced stress in Scots pine mimics, but does not entirely replicate, the effects of strong light stress [9]. Further research is required to elucidate the mechanisms of temperature stress across various plant species.

Additionally, the diversity of photosynthetic machinery organization and the primary steps of photosynthesis in vivo require further investigation. Recent findings have highlighted significant variations in energy transfer pathways and the structural organization of the photosynthetic machinery across different species. For example, initial results have revealed substantial differences in PSI organization between in vivo and in vitro studies of various plant species [36, 81]. This discovery suggests that photosynthetic mechanisms in plants are far more diverse than previously thought. Subsequent studies have uncovered even greater variation in PSI organization, underscoring the need for continued research in this area.

References

1. Chukhutsina VU, Buchel C, van Amerongen H (2013) Variations in the first steps of photosynthesis for the diatom Cyclotella meneghiniana grown under different light conditions. Biochim Biophys Acta Bioenerg 1827:10–18
2. Chukhutsina V, Bersanini L, Aro E-M, van Amerongen H (2015) Cyanobacterial flv4-2 operon-encoded proteins optimize light harvesting and charge separation in photosystem II. Mol Plant 8:747–761
3. Slavov C, Schrameyer V, Reus M, Ralph PJ, Hill R, Büchel C, Larkum AWD, Holzwarth AR (2016) "Super-quenching" state protects Symbiodinium from thermal stress – implications for coral bleaching. Biochim Biophys Acta Bioenerg 1857:840–847
4. Ünlü C, Drop B, Croce R, van Amerongen H (2014) State transitions in Chlamydomonas reinhardtii strongly modulate the functional size of photosystem II but not of photosystem I. Proc Natl Acad Sci USA 111:3460–3465
5. Van Stokkum IHM, Van Oort B, Van Mourik F, Gobets B, Van Amerongen H (2008) Biophysical techniques in photosynthesis. Springer, pp 223–240
6. Van Hoek A, Visser A (1985) Artefact and distortion sources in time correlated single photon counting. Instrum Sci Technol 14:359
7. Becker W (2012) Fluorescence lifetime imaging – techniques and applications. J Microsc 247: 119
8. Tian L, Xu P, Chukhutsina VU, Holzwarth ARR, Croce R (2017) Zeaxanthin-dependent nonphotochemical quenching does not occur in photosystem I in the higher plant Arabidopsis thaliana. Proc Natl Acad Sci USA 114:4828–4832
9. Bag P, Chukhutsina V, Zhang Z, Paul S, Ivanov AG, Shutova T, Croce R, Holzwarth AR, Jansson S (2020) Direct energy transfer from photosystem II to photosystem I confers winter sustainability in Scots Pine. Nat Commun 11:1–13

10. Holzwarth AR, Jahns P (2014) Demmig-Adams B, Garab G, Adams III W, Govindjee (eds) Non-photochemical quenching and energy dissipation in plants, algae and cyanobacteria. Springer, pp 129–156. https://doi.org/10.1007/978-94-017-9032-1_5
11. Chukhutsina VU, Fristedt R, Morosinotto T, Croce R (2017) Photoprotection strategies of the alga Nannochloropsis gaditana. Biochim Biophys Acta 1858:544–552
12. Le Quiniou C, Tian L, Drop B, Wientjes E, van Stokkum IH, van Oort B, Croce R (2015) PSI-LHCI of Chlamydomonas reinhardtii: increasing the absorption cross section without losing efficiency. Biochim Biophys Acta Bioenerg 1847:458–467
13. Chukhutsina V, Bersanini L, Aro E-M, Van Amerongen H (2015) Cyanobacterial light-harvesting phycobilisomes uncouple from photosystem I during dark-to-light transitions. Sci Rep 5:14193
14. Holzwarth AR, Miloslavina Y, Nilkens M, Jahns P (2009) Identification of two quenching sites active in the regulation of photosynthetic light-harvesting studied by time-resolved fluorescence. Chem Phys Lett 483:262–267
15. van Oort B, Alberts M, de Bianchi S, Dall'Osto L, Bassi R, Trinkunas G, Croce R, van Amerongen H (2010) Effect of antenna-depletion in photosystem II on excitation energy transfer in Arabidopsis thaliana. Biophys J 98:922–931
16. van der Weij-de Wit CD, Dekker JP, van Grondelle R, van Stokkum IH (2011) Charge separation is virtually irreversible in photosystem II core complexes with oxidized primary quinone acceptor. J Phys Chem A 115:3947–3956
17. Miloslavina Y, Szczepaniak M, Muller MG, Sander J, Nowaczyk M, Rogner M, Holzwarth AR (2006) Charge separation kinetics in intact photosystem II core particles is trap-limited. A picosecond fluorescence study. Biochemistry 45:2436–2442
18. Chukhutsina VU, Büchel C, Van Amerongen H, Buchel C, Van Amerongen H (2014) Disentangling two non-photochemical quenching processes in Cyclotella meneghiniana by spectrally-resolved picosecond fluorescence at 77K. Biochim Biophys Acta Bioenerg 1837: 899–907
19. Tian L, Gwizdala M, van Stokkum IHM, Koehorst RB, Kirilovsky D, van Amerongen H (2012) Picosecond kinetics of light harvesting and photoprotective quenching in wild-type and mutant phycobilisomes isolated from the cyanobacterium Synechocystis PCC 6803. Biophys J 102: 1692–1700
20. Farooq S, Chmeliov J, Wientjes E, Koehorst R, Bader A, Valkunas L, Trinkunas G, van Amerongen H (2018) Dynamic feedback of the photosystem II reaction centre on photoprotection in plants. Nat Plants 4:225–231
21. Chukhutsina VU, Holzwarth AR, Croce R (2019) Time-resolved fluorescence measurements on leaves: principles and recent developments. Photosynth Res 140:355–369
22. Elias E, Hu C, Croce R (2024) Inter-protein energy transfer dynamics in the PSII antenna. Cell Rep Phys Sci 5:102198
23. Barzda V, Gulbinas V, Kananavicius R, Cervinskas V, Van Amerongen H, Van Grondelle R, Valkunas L (2001) Singlet–singlet annihilation kinetics in aggregates and trimers of LHCII. Biophys J 80:2409–2421
24. Gruber JM, Chmeliov J, Kruger TP, Valkunas L, van Grondelle R (2015) Singlet-triplet annihilation in single LHCII complexes. Phys Chem Chem Phys 17:19844–19853
25. Van Grondelle R (1985) Excitation energy transfer, trapping and annihilation in photosynthetic systems. Biochim Biophys Acta Rev Bioenerg 811:147–195
26. Valkunas L, Trinkunas G, Liuolia V, van Grondelle R (1995) Nonlinear annihilation of excitations in photosynthetic systems. Biophys J 69:1117. https://doi.org/10.1016/S0006-3495(95)79986-6
27. Szczepaniak M, Sander J, Nowaczyk M, Muller MG, Rogner M, Holzwarth AR (2009) Charge separation, stabilization, and protein relaxation in photosystem II core particles with closed reaction center. Biophys J 96:621–631
28. Ahmad P, Prasad MNV (2012) Environmental adaptations and stress tolerance of plants in the era of climate change. Springer

29. Müller MG, Georgakoudi I, Zhang Q, Wu J, Feld MS (2001) Intrinsic fluorescence spectroscopy in turbid media: disentangling effects of scattering and absorption. Appl Opt 40:4633–4646
30. Patterson MS, Pogue BW (1994) Mathematical model for time-resolved and frequency-domain fluorescence spectroscopy in biological tissues. Appl Opt 33:1963–1974
31. Lakowicz JR (1994) Topics in fluorescence spectroscopy. Probe design and chemical sensing, vol 4. Springer Science & Business Media
32. Terjung F (1998) Reabsorption of chlorophyll fluorescence and its effects on the spectral distribution and the picosecond decay of higher plant leaves. Z Nat C 53:924–926
33. Connolly JS, Janzen AF, Samuel EB (1982) Fluorescence lifetimes of chlorophyll a: solvent, concentration and oxygen dependence. Photochem Photobiol 36:559–563
34. Mammel U, Brun M, Oelkrug D (1992) Absorption and emission spectroscopy in scattering media: modelling of fluorescence decay curves and photometric laws for systems with inhomogeneous absorber distribution. Fresenius J Anal Chem 344:147–152
35. Maxwell K, Johnson GN (2000) Chlorophyll fluorescence – a practical guide. J Exp Bot 51: 659–668
36. Chukhutsina VU, Liu X, Xu P, Croce R (2020) Light-harvesting complex II is an antenna of photosystem I in dark-adapted plants. Nat Plants 6:860–868
37. Gitelson AA, Buschmann C, Lichtenthaler HK (1998) Leaf chlorophyll fluorescence corrected for re-absorption by means of absorption and reflectance measurements. J Plant Physiol 152: 283–296
38. Agati G, Fusi F, Mazzinghi P, Dipaolo ML (1993) A simple approach to the evaluation of the reabsorption of chlorophyll fluorescence-spectra in intact leaves. J Photochem Photobiol B Biol 17:163–171
39. Lagorio MG, Dicelio LE, Litter MI, Roman ES (1998) Modeling of fluorescence quantum yields of supported dyes – aluminium carboxyphthalocyanine on cellulose. J Chem Soc Faraday Trans 94:419–425
40. Ramos ME, Lagorio MG (2004) True fluorescence spectra of leaves. Photochem Photobiol Sci 3:1063–1066
41. Pellegrino F (1983) Ultrafast energy transfer processes in photosynthetic systems probed by picosecond fluorescence spectroscopy. Opt Eng 22:225508
42. Schmuck G, Moya I (1994) Time-resolved chlorophyll fluorescence spectra of intact leaves. Remote Sens Environ 47:72–76
43. Briantais JM, Dacosta J, Goulas Y, Ducruet JM, Moya I (1996) Heat stress induces in leaves an increase of the minimum level of chlorophyll fluorescence, Fo: a time-resolved analysis. Photosynth Res 48:189–196
44. Miloslavina Y, de Bianchi S, Dall'Osto L, Bassi R, Holzwarth AR (2011) Quenching in Arabidopsis thaliana mutants lacking monomeric antenna proteins of photosystem II. J Biol Chem 286:36830–36840
45. Schreiber U (2004) Pulse-amplitude-modulation (PAM) fluorometry and saturation pulse method: an overview. In: Chlorophyll a fluorescence. Springer, pp 279–319
46. Joliot P, Joliot A (1984) Electron transfer between the two photosystems. I. Flash excitation under oxidizing conditions. Biochim Biophys Acta Bioenerg 765:210
47. Nanda S, Shutova T, Cainzos M, Hu C, Sasbrink B, Bag P, den Blanken T, Buijs R, van der Gracht L, Hendriks F (2024) ChloroSpec: a new in vivo chlorophyll fluorescence spectrometer for simultaneous wavelength-and time-resolved detection. Physiol Plant 176:e14306
48. Genty B, Briantais J-M, Baker NR (1989) The relationship between the quantum yield of photosynthetic electron transport and quenching of chlorophyll fluorescence. Biochim Biophys Acta Gen Subj 990:87–92
49. Cornic G, Ghashghaie J (1991) Effect of temperature on net CO 2 assimilation and photosystem II quantum yield of electron transfer of French bean (Phaseolus vulgaris L.) leaves during drought stress. Planta 185:255–260

50. Krall JP, Edwards GE (1990) Quantum yields of photosystem II electron transport and carbon dioxide fixation in C4 plants. Funct Plant Biol 17:579–588
51. Pfündel E (1998) Estimating the contribution of photosystem I to total leaf chlorophyll fluorescence. Photosynth Res 56:185–195
52. Murata N (1969) Control of excitation transfer in photosynthesis I. Light-induced change of chlorophyll a fluorescence in Porphyridium cruentum. Biochim Biophys Acta Bioenerg 172: 242–251
53. Miloslavina Y, Wehner A, Lambrev PH, Wientjes E, Reus M, Garab G, Croce R, Holzwarth AR (2008) Far-red fluorescence: a direct spectroscopic marker for LHCII oligomer formation in non-photochemical quenching. FEBS Lett 582:3625–3631
54. Bag P, Chukhutsina V, Zhang Z, Paul S, Ivanov A, Shutova T, Croce R, Holzwarth A, Jansson S (2020) Direct energy transfer from photosystem II to photosystem I is the major regulator of winter sustainability of Scots pine. Nat Commun 11:6388
55. van Stokkum IHM, Larsen DS, van Grondelle R (2004) Global and target analysis of time-resolved spectra. Biochim Biophys Acta Bioenerg 1657:82–104
56. Holzwarth AR (1995) Methods in enzymol, vol 246. Academic Press, pp 334–362
57. Krahl R, Bülter A, Koberling F (2005) Performance of the micro photon devices PDM 50CT SPAD detector with PicoQuant TCSPC systems. Technical Note. PicoQuant GmbH
58. Chukhutsina VU, Büchel C, van Amerongen H, Buchel C, van Amerongen H, Büchel C, van Amerongen H (2013) Variations in the first steps of photosynthesis for the diatom Cyclotella meneghiniana grown under different light conditions. Biochim Biophys Acta Bioenerg 1827: 10–18
59. Schiller NH (1984) Semiconductors probed by ultrafast laser spectroscopy. https://doi.org/10.1016/b978-0-12-049902-1.50018-9
60. Tian L, van Stokkum IHM, Koehorst RBM, Jongerius A, Kirilovsky D, van Amerongen H (2011) Site, rate, and mechanism of photoprotective quenching in cyanobacteria. J Am Chem Soc 133:18304–18311
61. Shibata Y, Nishi S, Kawakami K, Shen JR, Renger T (2013) Photosystem II does not possess a simple excitation energy funnel: time-resolved fluorescence spectroscopy meets theory. J Am Chem Soc 135:6903
62. Mekala NR, Suorsa M, Rantala M, Aro EM, Tikkanen M (2015) Plants actively avoid state transitions upon changes in light intensity: role of light-harvesting complex ii protein dephosphorylation in high light. Plant Physiol 168:721
63. Ruban AV, Johnson MP (2009) Dynamics of higher plant photosystem cross-section associated with state transitions. Photosynth Res 99:173
64. Biddington NL (1986) The effects of mechanically-induced stress in plants – a review. Plant Growth Regul 4:103–123
65. Chen Z, Li J, Wang B-C, Tian L (2025) In vivo two-photon FLIM resolves photosynthetic properties of maize bundle sheath cells. Photosynth Res 163:1–9
66. Nozue S, Mukuno A, Tsuda Y, Shiina T, Terazima M, Kumazaki S (2016) Characterization of thylakoid membrane in a heterocystous cyanobacterium and green alga with dual-detector fluorescence lifetime imaging microscopy with a systematic change of incident laser power. Biochim Biophys Acta Bioenerg 1857:46
67. Broess K, Borst JW, Van Amerongen H (2009) Applying two-photon excitation fluorescence lifetime imaging microscopy to study photosynthesis in plant leaves. Photosynth Res 100:89
68. Yuan X, Bowling A, Gemperline E, Mannam V, Howard S (2024) Real-time, depth-resolved, in vivo multiphoton fluorescence lifetime imaging microscopy of agricultural herbicide treatments in plants. Opt Express 32:13733–13745
69. Göppert-Mayer M (1931) Über Elementarakte mit zwei Quantensprüngen. Ann Phys 401:273
70. Piston DW (1999) Imaging living cells and tissues by two-photon excitation microscopy. Trends Cell Biol 9:66. https://doi.org/10.1016/S0962-8924(98)01432-9
71. Croce R (2015) PsbS is the plants' pick for sun protection. Nat Struct Mol Biol 22. https://doi.org/10.1038/nsmb.3079

72. Niyogi KK, Grossman AR, Björkman O (1998) Arabidopsis mutants define a central role for the xanthophyll cycle in the regulation of photosynthetic energy conversion. Plant Cell 10:1121
73. Wientjes E, Croce R (2012) PMS: photosystem I electron donor or fluorescence quencher. Photosynth Res 111:185–191
74. Ballottari M, Alcocer MJP, D'Andrea C, Viola D, Ahn TK, Petrozza A, Polli D, Fleming GR, Cerullo G, Bassi R (2014) Regulation of photosystem I light harvesting by zeaxanthin. Proc Natl Acad Sci USA 111:E2431
75. Iermak I, Vink J, Bader AN, Wientjes E, Van Amerongen H (2016) Visualizing heterogeneity of photosynthetic properties of plant leaves with two-photon fluorescence lifetime imaging microscopy. Biochim Biophys Acta Bioenerg 1857:1473
76. Wientjes E, Philippi J, Borst JW, van Amerongen H (2017) Imaging the photosystem I/photosystem II chlorophyll ratio inside the leaf. Biochim Biophys Acta Bioenerg 1858:259–265
77. Verhoeven D, van Amerongen H, Wientjes E (2023) Single chloroplast in folio imaging sheds light on photosystem energy redistribution during state transitions. Plant Physiol 191:1186
78. Tikkanen M, Grieco M, Aro EM (2011) Novel insights into plant light-harvesting complex II phosphorylation and 'state transitions'. Trends Plant Sci 16:126–131
79. Pan X, Ma J, Su X, Cao P, Chang W, Liu Z, Zhang X, Li M (2018) Structure of the maize photosystem I supercomplex with light-harvesting complexes I and II. Science 1979(360): 1109–1113
80. Akhtar P, Lingvay M, Kiss T, Deák R, Bóta A, Ughy B, Garab G, Lambrev PH (2016) Excitation energy transfer between light-harvesting complex II and photosystem I in reconstituted membranes. Biochim Biophys Acta Bioenerg 1857:462–472
81. Bos PR, Schiphorst C, Kercher I, Buis S, de Jong D, Vunderink I, Wientjes E (2023) Spectral diversity of photosystem I from flowering plants. Photosynth Res 155:35

Proximal Remote Sensing: Simultaneous Active and Passive Measurements of Fluorescence at the Interface from Single Leaves to the Canopy

Juan Manuel Romero, Deepthi Konche, Sofia Choza-Farias, Bastian Siegmann, Oliver Knopf, Francisco Pinto, Uwe Rascher, and Onno Muller

Contents

J. M. Romero (✉)
Institute of Bio- and Geosciences 2 (IBG-2), Plant Sciences, Forschungszentrum Jülich, Jülich, Germany

Departamento de Química Inorgánica, Analítica y Química Física, Facultad de Ciencias Exactas y Naturales, Universidad de Buenos Aires, Buenos Aires, Argentina

CONICET, Instituto de Investigaciones Fisiológicas y Ecológicas vinculadas a la Agricultura (IFEVA), Buenos Aires, Argentina

D. Konche, S. Choza-Farias, B. Siegmann, O. Knopf, and O. Muller
Institute of Bio- and Geosciences 2 (IBG-2), Plant Sciences, Forschungszentrum Jülich, Jülich, Germany
e-mail: d.konche@fz-juelich.de; s.choza.farias@fz-juelich.de; b.siegmann@fz-juelich.de; o.muller@fz-juelich.de

F. Pinto
Department of Plant Sciences, Centre for Crop Systems Analysis, Wageningen University Research, Wageningen, The Netherlands
e-mail: francisco.pinto@wur.nl

U. Rascher
Institute of Bio- and Geosciences 2 (IBG-2), Plant Sciences, Forschungszentrum Jülich, Jülich, Germany

Faculty of Agricultural, Nutritional and Engineering Sciences, Institute of Crop Science and Resource Conservation, University of Bonn, Bonn, Germany
e-mail: u.rascher@fz-juelich.de

M. G. Lagorio (ed.), Comprehensive Series in Photochemical and Photobiological Sciences, Fluorescence of Living Plants 1, https://doi.org/10.1007/17361_2025_18
Published online: 26 August 2025

Abstract Detailed physiological information from crops is essential for various applications, including breeding, farming, and ecological assessment. Ideally, such information is obtained through noninvasive techniques, among which visible spectroscopy is the most widely used. Although reflectance measurements provide reliable information, chlorophyll fluorescence (ChlF) is considered the most promising physiological indicator due to its close association with photosystem II chemistry.

ChlF can be retrieved at multiple spatial scales, from chloroplasts and individual leaves to farm plots, entire fields, and even regional scales. Notably, in remote sensing applications, ChlF is primarily retrieved passively, whereas at the leaf level, it is commonly measured actively. Therefore, it is crucial to bridge both retrieval approaches through simultaneous measurements at the canopy scale, which serves as the intermediate level.

In this study, we demonstrate how a combined active and passive measurement approach can provide new insights into the physiology of two wheat varieties subjected to drought conditions. Photosynthetic efficiency and electron transport rate (ETR) were estimated using active fluorescence measurements with a light-induced fluorescence transient (LIFT) sensor. Passive measurements were conducted with a fluorescence box device (FloX), equipped with two spectrometers, enabling the calculation of reflectance-based vegetation indices and the retrieval of sun-induced fluorescence (SIF) via fluorescence line discrimination (FLD) methods. Both devices were mounted on a field cart to allow systematic and reproducible measurements throughout the growing season.

This integrated approach revealed a strong relationship between SIF emission and ETR, both of which are linked to total absorbed photosynthetically active radiation (APAR). We also observed that while drought conditions significantly reduced photosynthetic efficiency over the season, SIF yield remained relatively constant. Consequently, no correlation between photosynthetic and fluorescence emission yields was detected. This behavior suggests that, in this case study, fluorescence emission does not appear to be the primary mechanism by which plants mitigate photooxidative stress.

The approach presented here offers a framework for future simultaneous retrieval of fluorescence at the canopy level and lays the foundation for interpreting remote SIF measurements from both physiological and photochemical perspectives.

Keywords Chlorophyll fluorescence (ChlF), Drought stress, Photosynthetic efficiency, Remote sensing, Sun-induced fluorescence (SIF)

1 Motivation

Monitoring of vegetation by means of noninvasive techniques is one of the main interests not only for the scientific community but also for several disciplines. Breeders pursue the understanding of genotype, environment, and management (G × E × M) interactions for improving crop productivity by selecting heritable traits [1, 2]. Farmers, governments, and companies aim to predict the yield of crops and detect diseases with enough anticipation to take duly informed agronomic decisions [3]. Moreover, ecosystems and climate scientists need remote sensing tools that allow the estimation of biomass production and carbon uptake by forests and other vegetation covers, with large spatial and time coverage [4].

The spatial scale at which the information is retrieved is highly dependent on the application, ranging from the centimeter-meter scale for phenotyping to kilometers for agronomic monitoring and earth observation [5]. This wide variety of spatial scales calls for the understanding of the physical, physiological, and structural processes affecting measurements at leaf, canopy, and remote scales [6, 7].

As stated in other chapters in this book, leaf optics have been widely studied since the 1970s [8]. Reflectance and transmittance have been used, for example, for pigment content estimation [9] and for the early detection of damage caused by herbicides [10]. Given its close link to the light reactions of photosynthesis, chlorophyll fluorescence has proven to be an additional, sometimes more sensitive way to monitor environmental effects that affect the functional status of plants [11]. For example, variable fluorescence that is determined by active measurement approaches has been shown to reveal the effect of temperature increase [12], drought [13], heavy metals [14], and several pesticides [15, 16].

Stationary fluorescence that can also be retrieved on the larger scale has gained increasing attention in the past years and has been deeply studied from both a physiological and a physicochemical perspective [17–19], yielding interesting studies showing, for example, the change in the spectral distribution of fluorescence in adaxial and abaxial faces due to different light environments [20], the effect of different heavy metals on photosynthesis [21, 22], or the impact of air pollution in urban trees [23]. On the large spatial scale, reflectance and sun-induced fluorescence (SIF) are retrieved from satellite measurements and used to monitor primary productivity [24, 25], radiation use efficiency [26], and impact of phosphate limitation on plant productivity [27]. The selection of FLEX to become the eighth Earth Explorer satellite mission of ESA [28] has significantly increased interest in understanding dynamic changes in vegetation and their link to plant photosynthesis at larger spatial scales [29], and nowadays SIF data can be retrieved from monitoring towers [30, 31], airplanes [32–34], and satellites [35–37]. In 2026, global SIF data will be greatly extended by launching the dedicated FLEX satellite mission, which will provide global fluorescence and reflectance data with high spatial resolution, narrowing the gap between experimental field crop research and remote sensing [38, 39].

As it may arise from this brief characterization, plant monitoring at the leaf scale has been mostly focused on active measurements of variable, spectrally integrated fluorescence, whereas remote sensing applications rely almost totally on passive retrieval of SIF, which is a stationary signal. This is mainly due to the fact that although active variable fluorescence is the most sensitive proxy for physiological assessment, the technical limitations to retrieve it in the field are immense; thus, SIF is the only available signal for levels higher than the canopy scale. Therefore, there is a great need to close the gap between these two different but complementary types of measurements at the interphase: the canopy scale.

In this chapter, we show the potential of the simultaneous retrieval of both active and passive fluorescence at the canopy level in the field. The conceived work scheme was applied as an example to a general case study, understanding the effect of progressive drought in a wheat breeding trail. We provide a detailed description of the protocol and of the possible resulting outcomes.

2 What Is Canopy Photosynthesis and Canopy Fluorescence?

Canopy photosynthesis can be defined and measured in several different ways, and the best option should be chosen taking into account the question that aims to be answered. Gross primary production (GPP) is the total amount of CO_2 assimilated by the canopy, and it is usually measured by means of eddy covariance towers [40]. It depends not only on the amount of energy harvested from the sun and used in the electron transport chain (that is, ETR) but also on the availability of CO_2 and the efficiency of the Calvin cycle.

Electron transport rate (ETR) is the amount of energy flowing through the PSII and reflects the conversion of light energy into chemical energy, an electron flow. It is retrieved from variable fluorescence experiments, mostly at leaf scale [41]. It depends mainly on the amount of available photosystems, the incident radiation flux, and the regulatory mechanisms that couple/decouple the antenna complexes from the reaction centers or quench the chlorophyll excited state.

The relationship between ETR and GPP at the canopy scale is in principle proportional as they are both strongly linked to the total absorbed photosynthetically active radiation (APAR): the more electrons flow through the photosystems, the more reductive molecules are synthesized (NADPH) and more CO_2 molecules are reduced in the Calvin cycle. Although there are still no canopy ETR estimations – this work is a pioneer in this sense – studies that relate canopy and leaf measurements may support this idea [42]. However, several different factors can affect this relationship as, for example, sudden or fast changes in irradiance [43, 44] or pollution might have an immediate effect on ETR but not on GPP.

3 How Can We Measure Canopy Fluorescence?

Continuous measurements of variable fluorescence leading mainly to OJIP curves are widely spread for physiology monitoring at leaf level in controlled conditions [16]. Due to the intense ambient light conditions in crop fields, pulse amplitude modulated (PAM) fluorometry is required, as it effectively separates actively induced fluorescence signals from external spurious light [45].

To conduct individual measurements on several leaves in the same day, Mini-PAM is one of the most popular devices, as it allows rapid assessments of photosynthetic efficiency by measuring parameters such as F_v/F_m and ΦPSII in various plant species and ecosystems [46]. Moni-PAM, designed for long-term, unattended monitoring, continuously records chlorophyll fluorescence and temperature on the same leaf area over extended periods. It has been used to study the diurnal and seasonal acclimation of PSII in Scots pine canopies and in large-scale ecological projects like NASA's ABoVE to monitor photosynthetic activity in boreal forests [47].

The fast repetition rate technique for measuring variable chlorophyll fluorescence was presented some decades ago [48], but it was not until a few years ago that it became available as a prototype to measure variable fluorescence in the field [49]. The light-induced fluorescence transient (LIFT) method allows the estimation of photosynthetic rates at the canopy level with high time resolution under natural environmental conditions [50]. However, until now, no attempts have been made in order to link the physiological parameters retrieved by means of the LIFT method with the SIF signal measured by field spectrometers in crops.

For passive SIF measurements, the FloX system (JB Hyperspectral Devices, Düsseldorf, Germany) is capable of measuring over larger areas (ca. 1 m^2) and from distances up to 100 m [51], making it more suitable for broader canopy-level assessments [52, 53]. The FloX system uses sunlight as an excitation source and measures the plants' radiance by two spectrometers: a FULL module with a bandwidth from 400 to 950 nm and a resolution of 0.65 nm and a FLUO module with a bandwidth of 650–800 nm and a resolution of 0.17 nm. The combination of these two spectrometers allows the efficient retrieval of fluorescence through Fraunhofer line discrimination algorithms [54].

4 Case Study: Understanding Drought Effect in a Wheat Breeding Trail

Drought is one of the most significant climatic stresses that vegetation must tolerate, especially in the context of climate change [55]. Under relatively dry weather, drought is a progressive, manageable stress, therefore suitable for testing the sensitivity of the combined fluorescence retrieval approach. Interestingly, two independent studies showed that drought events lead to a double response in the fluorescence signal characterized by an initial signal increase followed by a decrease after a few days, which was demonstrated for wheat using active [56] and maize using passive fluorescence measurements [57]. Detecting the double response in the fluorescence signal is solely possible from temporal high-resolution data consisting of daily measurements covering the drought or heat event.

4.1 Trial Setup

A field experiment was conducted during the 2022–2023 field season at the CIMMYT's Norman E. Borlaug experimental station (CENEB) in Ciudad Obregon, Sonora, Mexico (27°23′46″N, 109°55′42″W, 38 m a.s.l.). The experiment consisted of a non-randomized factorial design, with two spring wheat genotypes and two irrigation treatments arranged in a structured field layout. Each combination was replicated four times, with genotypes assigned to adjacent plots within rows, and irrigation treatments assigned by blocks within columns. Two CIMMYT genotypes were selected based on their contrasting response to drought. The cross WBLL4//OAX93.24.35/WBLL1 was identified as a tolerant genotype and the cross PAVLOVKA/V15.89C//NAVJ07/3/ROLF07/4/MUCUY as a susceptible one. Water was supplied with standard drip irrigation (i.e., 4 L/h/emitter). Both treatments were grown under full irrigation (i.e., irrigation every ~10 days) until anthesis, where the water was withheld for the drought treatment whereas the irrigation frequency continued for the fully irrigated treatment (control, soil kept wet). The experiment was sown on November 16, 2022, and 50% emerged on November 30, 2022. The experiment was harvested on May 8, 2023.

As shown in Fig. 1, Panel a, the continuous bed of the experiments was measured with a field cart implemented with a light-induced fluorescence transient method (LIFT, active measurement of fluorescence) and a FloX (passive measurement of solar-induced fluorescence) targeting at a frequency of one measurement per day (depending on weather conditions) during the period between canopy closure and heading (a critical moment for stress adaptation). The measurements were conducted around solar noon (±1 h).

4.2 FloX Protocol

High-spectral-resolution passive fluorescence measurements at the canopy level were obtained using a fluorescence box (FloX; https://www.jb-hyperspectral.com/products/flox/ Dusseldorf, Germany). FloX contains two spectrometers: a QE Pro (Ocean Optics Inc., Dunedin, USA) with a spectral resolution of approximately 0.3 nm (SSI ≈ 0.15 nm) covering the 650–800 nm range for precise SIF detection and a FLAME S spectrometer (Ocean Optics, USA) covering the full range from 400 to 950 nm, enabling the computation of reflectance and vegetation indexes (VIs)

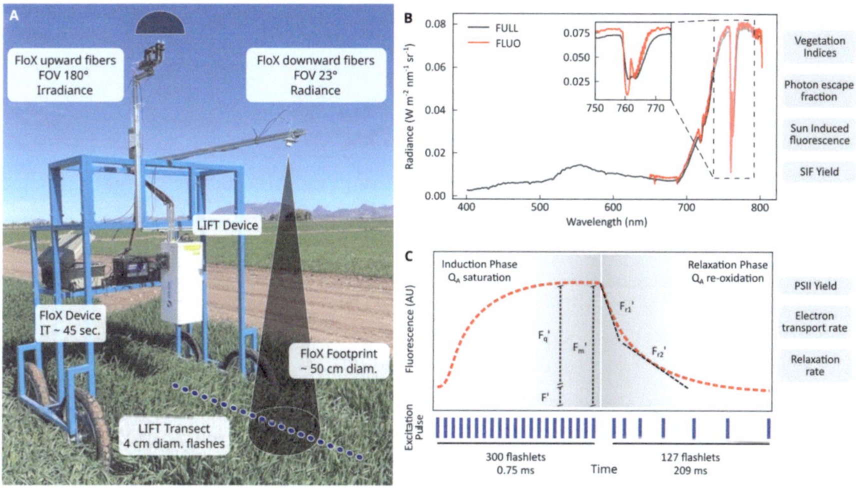

Fig. 1 Measuring methodology. Panel **a**: Devices and cart in measuring position. FloX footprint and LIFT transect are shown. *IT* FloX integration time. Panel **b**: FloX raw data. Raw spectra from FULL (black line) and FLUO (red line) spectrometers. Inset shows detail of the region used to retrieve SIF via FLD algorithms. Relevant information retrieved from FloX measurements shown in grey boxes on the right. Panel **c**: LIFT raw data. Example active fluorescence trace (red line, not in scale) and excitation pulse scheme (not in scale). F' minimum stationary fluorescence, F'_m maximum fluorescence in light-saturating conditions, F'_q variable fluorescence, F'_q/F'_m PSII operating efficiency, F'_{r1} *and* F'_{r2} recovery slopes 1 and 2, respectively. Relevant information retrieved from LIFT measurements shown in grey boxes on the right. (Panel **c** based on [58])

(see Fig. 1, Panel b). Each spectrometer captured the downwelling and upwelling radiance through a pair of optical fibers (see Fig. 1, Panel a).

From these reflectance data, VIs were derived. The normalized difference vegetation index (NDVI) was calculated as proposed by Tucker [59], and SIF was calculated using an improved Fraunhofer line discrimination method (iFLD [60]), as it provides the most reliable values [54]. iFLD uses plant cover radiance measurements (L) and solar irradiation (E) inside (in) and outside (out) the 760 nm band. SIF was obtained according to the following equation:

$$\mathrm{SIF} = \frac{\alpha_{\mathrm{R}} \cdot E_{\mathrm{out}} \cdot E_{\mathrm{in}} - L_{\mathrm{out}} \cdot E_{\mathrm{in}}}{\alpha_{\mathrm{R}} \cdot E_{\mathrm{out}} - \alpha_{\mathrm{F}} \cdot E_{\mathrm{in}}}, \tag{1}$$

where α_{R} and α_{F} indicate the reflectance and fluorescence correction factors, respectively, which account for the nonlinear variation of R and F inside and outside the absorption feature and for using the apparent reflectance (Rapp) instead of actual reflectance in the measurement:

$$\alpha_R = \frac{Rapp_{out}}{Rapp_{in}}, \tag{2}$$

$$\alpha_F = \alpha_R \cdot \frac{E_{out}}{E_{in}}. \tag{3}$$

$Rapp_{in}$ and E_{in} are calculated by nonlinear interpolation of reflectance and irradiance spectrum, respectively, by means of a cubic spline interpolation method.

Fluorescence emitted by vegetation and arriving at the detecting device over the canopy can be expressed as

$$\text{SIF} = \text{APAR} \cdot \text{SIF}_{\text{Yield}} \cdot f_{\text{esc}}, \tag{4}$$

where APAR is the absorbed photosynthetically active radiation, $\text{SIF}_{\text{Yield}}$ is the quantum yield of the fluorescence emission (emitted photon per absorbed photon), and f_{esc} is the probability of an emitted photon escaping the canopy and reaching the sensor.

To calculate APAR, the difference between incoming and reflected radiance was multiplied by the fraction of light absorbed by the vegetation, which was estimated as a function of NDVI following the approach of Moreau and Li [61]. The fraction of absorbed photosynthetically active radiation (fAPAR) was subsequently obtained by dividing APAR by incoming PAR.

Photon escape fraction was calculated according to the approach proposed by Zeng et al. [62], which is valid for the far-red region of the spectrum:

$$f_{\text{esc}} = \text{NDVI} \cdot R_{\text{IR}} \cdot \text{fAPAR}. \tag{5}$$

Last, $\text{SIF}_{\text{Yield}}$ was calculated following Mattera et al. [63] as the quotient:

$$\text{SIF}_{\text{Yield}} = \frac{\text{SIF}}{\text{APAR} \cdot f_{\text{esc}}}. \tag{6}$$

4.3 LIFT Protocol

A light-induced fluorescence transient device (model LIFT-REM 1.0, Soliense Inc., Shoreham, NY, USA; https://soliense.com/LIFT_Terrestrial.php) was employed to monitor variations in chlorophyll fluorescence (ChlF). The LIFT instrument was installed at the front end of a cart (model based on [64]) in the vertical direction pointing downward (nadir) and above the plant canopy, perpendicular to the crop row. The distance from the LIFT lens to the median uppermost canopy (i.e., the target area) was ~0.60 m, being regularly adjusted as plants grew. The blue LED light beam was ~0.04 m in diameter at the focal point over the target area. The fluorescence channel's gain was adjusted at the beginning of each measurement day to maintain the raw fluorescence within the 2,000–20,000 signal range, aiming to operate with an optimal signal-to-noise ratio.

The active probing method, first described by Kolber et al. [48], induces chlorophyll fluorescence by emitting a series of sub-saturating excitation light impulses at a fast repetition rate (FRR). By the FRR method, the capacity of electron transport to photosystem II (PSII) is exceeded, causing reaction centers to close. Therefore, resulting changes in the plants' fluorescence signal can be measured within microseconds by an avalanche photodiode in combination with an optical interference filter (685 nm ± 10 nm) that separates the red ChlF from the rest of the light.

Based on a slight modification of the procedure reported by Keller et al. [50], we derived the photosynthetic traits from the LIFT-measured ChlF transients using the QA (primary quinone electron acceptor) flash protocol. The PSII operating efficiency from light-adapted plants was estimated as the maximum ChlF yield observed between flashlets 298th and 302nd inclusive. During the initial excitation flashlets, most of the absorbed energy is utilized for photosynthetic activity, resulting in a low fluorescence signal observed during the first of these flashlets, referred to as minimum fluorescence (F'). As subsequent flashlets were applied, the fluorescence signal increased, reaching a maximum when most photosynthetic reaction centers were closed. This saturation phase (see Fig. 1, Panel c) typically involved around 300 flashlets, applied at 2.5 μs intervals, lasting approximately 0.75 ms, yielding the maximum fluorescence yield (F'_m). The difference between these fluorescence yields resulted in variable fluorescence (F'_q), which was used to calculate PSII yield (F'_q/F'_m) [50]. This protocol also includes a relaxation phase that lasts around 209 ms and consists of 127 flashlets separated by exponentially increasing times. This phase provides the recovery slopes 1 and 2, which are the kinetic parameters that characterize QA reoxidation (Fig. 1, Panel c; not analyzed in this study).

LIFT operated in a continuous scanning mode along an extended scan line in each plot, enabling measurements from multiple sun-exposed leaves within the canopy. The scans were performed at a speed of about 10 cm s^{-1} along the middle row of each plot (1.6 × 5 m), with 50 data points collected in each plot, each representing a different leaf. To quantify canopy photosynthesis, electron transport rate (ETR) was calculated as follows:

$$\mathrm{ETR} = \mathrm{PSII}_{\mathrm{Yield}} \cdot \mathrm{Irrad} \cdot \beta \tag{7}$$

where

- $\mathrm{PSII}_{\mathrm{Yield}}$ = Effective quantum yield of PSII (F'_q/F'_m)
- Irrad = Incident photosynthetically active radiation (measured by LIFT)
- β = Fraction of absorbed light allocated to PSII (typically assumed to be 0.5)

5 Results

5.1 Reflectance-Based Vegetation Indices

Figure 2 shows the time progression of the calculated vegetation indices along the season. As seen in Panel a, normalized difference vegetation index (NDVI) tracks the greenness decrease upon drought effect in a sensitive way. Stable, statistically significant decreases upon non-irrigation treatment are observed from February 8 for the susceptible variety and from February 14 for the tolerant variety (Northern

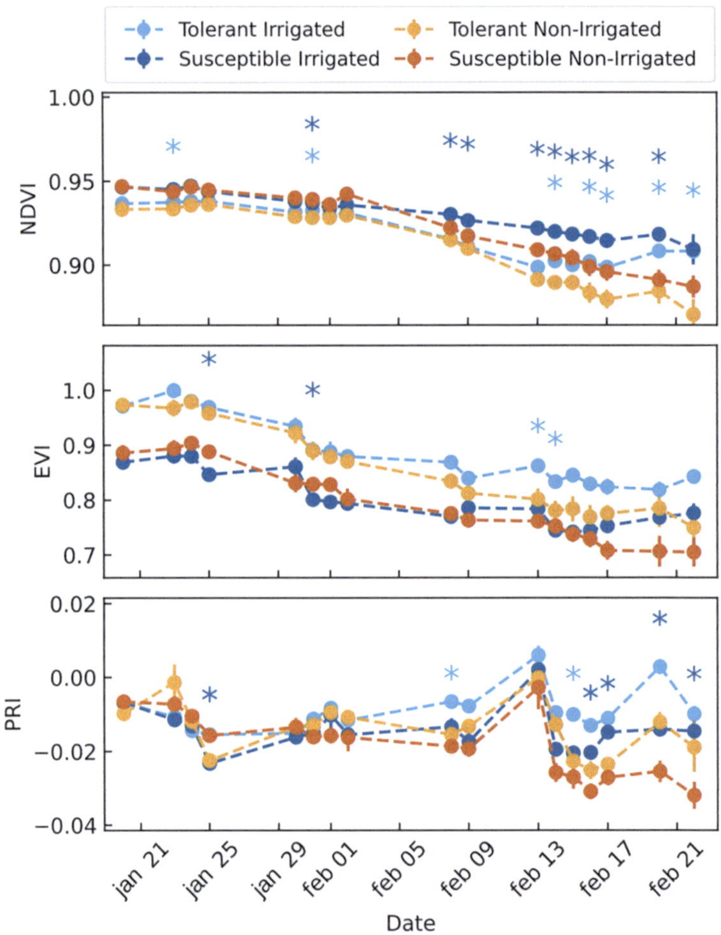

Fig. 2 Seasonal dynamics of reflectance-based phenology indicators. Normalized difference vegetation index (NDVI, Panel **a**), enhanced vegetation index (NDVI, Panel **b**), and photochemical reflectance index (PRI, Panel **c**) for tolerant variety in irrigated and nonirrigated conditions (light blue and light orange, respectively) and susceptible variety in irrigated and nonirrigated conditions (dark blue and dark orange, respectively). Error bars indicate the mean square error (MSE). Asterisks indicate significant differences between irrigated and nonirrigated conditions ($p < 0.05$)

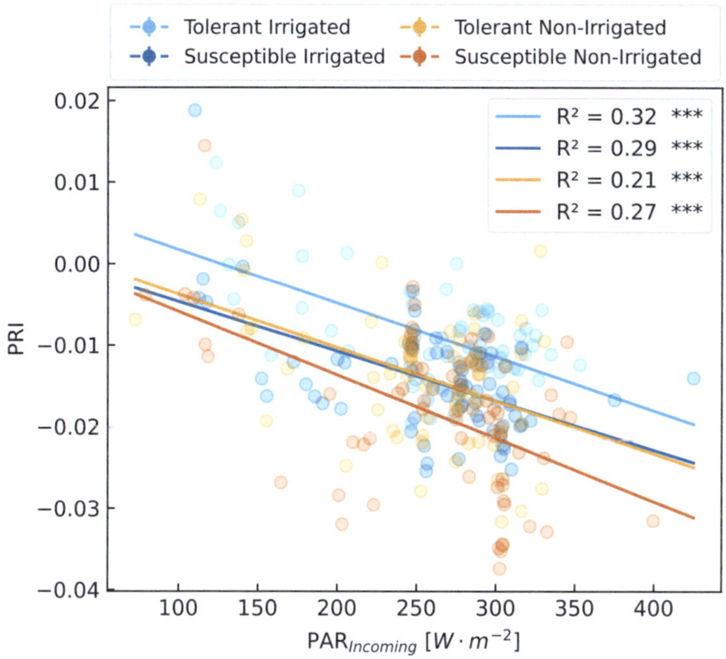

Fig. 3 Relationship between photochemical reflectance index (PRI) and incoming PAR for tolerant variety in irrigated and nonirrigated conditions (light blue and light orange, respectively) and susceptible variety in irrigated and nonirrigated conditions (dark blue and dark orange, respectively). Linear regression lines were fitted using the least squares method. R^2 values are shown in the plots. Asterisks denote the statistical significance of the fit: * $p < 0.1$; ** $p < 0.05$; *** $p < 0.01$

Hemisphere winter). Enhanced vegetation index (EVI [65]), although generally more sensitive than NDVI to greenness changes, does not show the effect of drought. Interestingly, EVI reveals an increased green biomass for the tolerant variety when compared to the susceptible one throughout the season. Last, PRI shows no effect of drought. As shown in Fig. 3, this is because PRI is much more related to light stress than limitations in soil water availability.

5.2 *Energy Fluxes*

Figure 4 shows the time course of the three main energy fluxes during the season: absorbed energy (APAR), energy used for photosynthesis (ETR), and energy released as fluorescence in the far red (SIF). Figure 4, Panel b, shows how the trend in the total energy flowing into the electron transport chain (ETR) is highly similar to the one of APAR (see also Fig. 5, Panel a). However, ETR reveals a statistically significant decrease for the susceptible variety from February 13. SIF (Panel c), on the other hand, does not reveal significant differences as a result of the

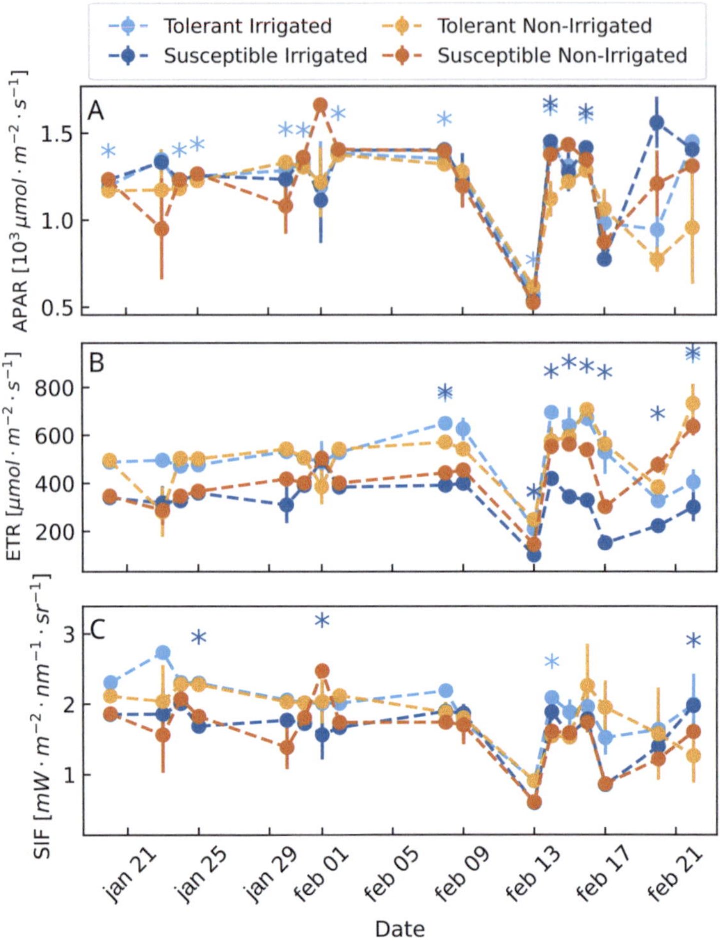

Fig. 4 Seasonal dynamics of energy fluxes. Absorbed photosynthetically active (APAR, Panel **a**), electron transport rate of PSII (ETR, Panel **b**), and far-red sun-induced fluorescence (SIF, Panel **c**) for tolerant variety in irrigated and nonirrigated conditions (light blue and light orange, respectively) and susceptible variety in irrigated and nonirrigated conditions (dark blue and dark orange, respectively). Error bars indicate the mean square error (MSE). Asterisks indicate significant differences between irrigated and nonirrigated conditions ($p < 0.05$)

irrigation treatments. Interestingly, in this experiment, the main driver of the total absorbed radiation is the incoming PAR rather than the fraction absorbed by the vegetation (Fig. 5).

Figure 6 illustrates the correlation between energy fluxes for all the plots measured for each treatment during the growing season. As expected, both canopy ETR and SIF are positively related to the total amount of energy absorbed (Panels a and b, respectively). This strong dependence on APAR is the main driver coupling ETR and SIF (Panel c), highlighting the potential of SIF as a proxy for canopy photosynthesis.

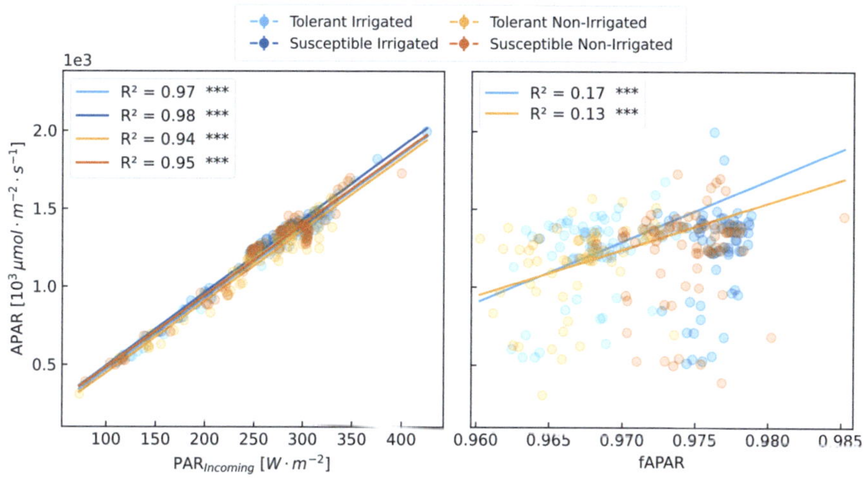

Fig. 5 Relationship between APAR and PAR (Panel **a**) or fAPAR (Panel **b**) for tolerant variety in irrigated and nonirrigated conditions (light blue and light orange, respectively) and susceptible variety in irrigated and nonirrigated conditions (dark blue and dark orange, respectively). Linear regression lines were fitted using the least squares method. R^2 values are shown in the plots. Asterisks denote the statistical significance of the fit: * $p < 0.1$; ** $p < 0.05$; *** $p < 0.01$

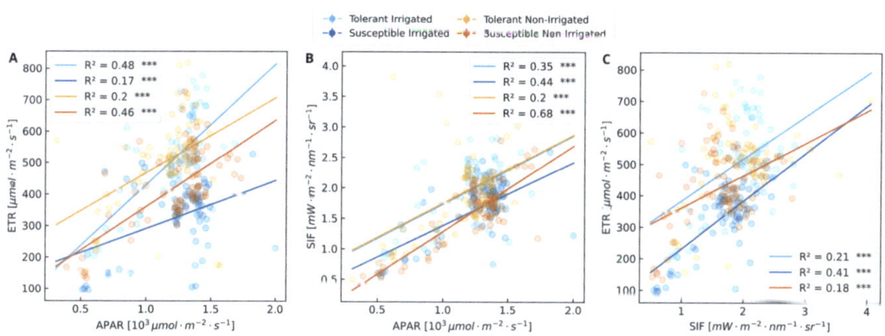

Fig. 6 Relationship between energy fluxes for tolerant variety in irrigated and nonirrigated conditions (light blue and light orange, respectively) and susceptible variety in irrigated and nonirrigated conditions (dark blue and dark orange, respectively). Panel **a** shows the correlation between ETR and APAR, while Panel **b** shows the correlation between SIF and APAR. Last, Panel **c** shows the correlation between ETR and SIF. Linear regression lines were fitted using the least squares method. R^2 values are shown in the plots. Asterisks denote the statistical significance of the fit: * $p < 0.1$; ** $p < 0.05$; *** $p < 0.01$

5.3 *Photosynthesis and Fluorescence Yields*

Panel a in Fig. 7 shows the time course of the incoming energy where measurements on cloudy days are easily identified (February 13 and 17). Although fluctuations in PAR cause small fluctuations in PSII yield, the quantum efficiency of PSII is

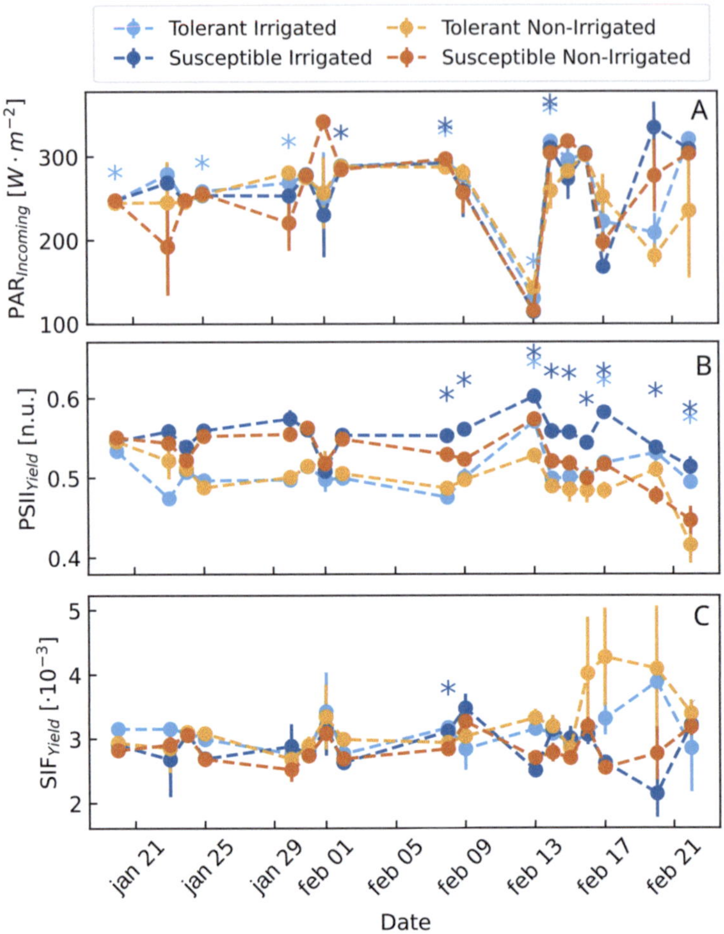

Fig. 7 Seasonal dynamics of PAR and energy dissipation yields. Absorbed photosynthetically active (APAR, Panel **a**), electron transport rate of PSII (ETR, Panel **b**), and far-red sun-induced fluorescence (SIF, Panel **c**) for tolerant variety in irrigated and nonirrigated conditions (light blue and light orange, respectively) and susceptible variety in irrigated and nonirrigated conditions (dark blue and dark orange, respectively). Error bars indicate the mean square error (MSE). Asterisks indicate significant differences between irrigated and nonirrigated conditions ($p < 0.05$)

retrieved by the LIFT protocol in a stable and reproducible way (Panel b). As with NDVI, significant differences due to the drought effect are observed from February 8 for the susceptible variety and from February 14 for the tolerant variety. However, despite the fact that in stressed plants less energy is going through the electron transport chain, no significant increase of fluorescence emission yields is observed (Panel c). This leads to a general lack of correlation between PSII and SIF yields along this season (Fig. 8).

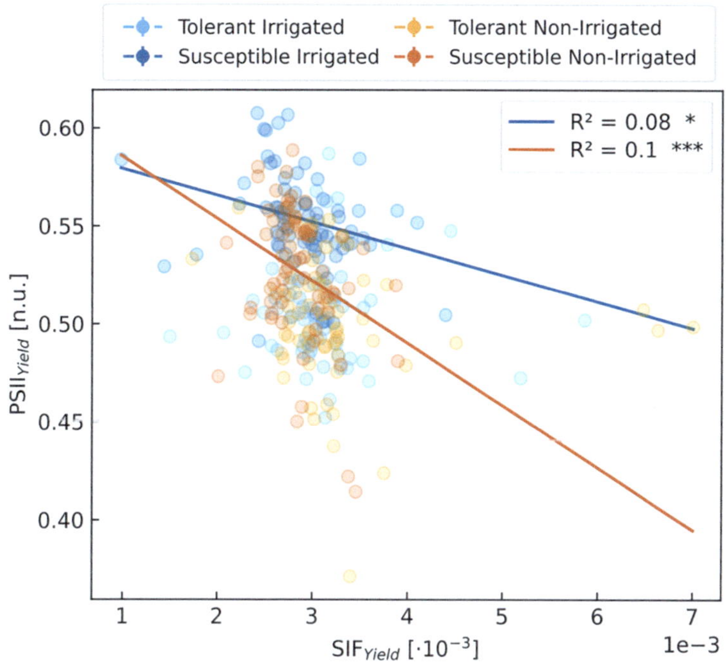

Fig. 8 Relationship between PSII yield and SIF yield for tolerant variety in irrigated and nonirrigated conditions (light blue and light orange, respectively) and susceptible variety in irrigated and nonirrigated conditions (dark blue and dark orange, respectively). Linear regression lines were fitted using the least squares method. R^2 values are shown in the plots. Asterisks denote the statistical significance of the fit: * $p < 0.1$; ** $p < 0.05$; *** $p < 0.01$

6 Discussion

In this study, we present for the first time the combined approach of active and passive fluorescence measurements at the canopy scale, the interphase between leaf and remote measurements, identified as proximal remote sensing (Fig. 1). The question chosen to test this approach is how two different wheat varieties react to an increasing drought stress. We hypothesized that the implemented approach allows linking ETR, a parameter generally retrieved at leaf level, with SIF information, usually measured in a remote manner.

In agreement with previous studies [66], analysis of reflectance-based spectral indices shows that NDVI tracks green biomass changes upon drought in a sensitive way, showing a significant decrease in nonirrigated plants on February 8 for the susceptible variety and 1 week later for the tolerant variety (*S*). Furthermore, our results indicate that NDVI is more sensitive than EVI for detecting the impact of drought on wheat biomass (Fig. 2). EVI was designed not to saturate with high LAI values but shows generally a smaller dynamic range than NDVI [67]. Although the comparison between NDVI and EVI was not conducted over a drought experiment on this type of crop, the smaller dynamic range observed for EVI might explain the latter's inability to early detect drought effects. Last, PRI showed no link with structural changes upon drought but instead was significantly negatively correlated with incident PAR for all the plots and treatments (Fig. 3). Though PRI could be an indicator of water stress at leaf level under constant light intensity conditions [68], this is not the case for PRI at the canopy level under field conditions, as the index is much more affected by the PAR fluctuations (Fig. 3).

The light energy absorbed by plants can follow one of the three pathways: electron transport chain (leading to photosynthesis), fluorescence emission, and dissipation as heat. Although the latter cannot be measured at canopy level, analyzing the dynamics of ETR and SIF leads to the understanding of how plants absorb energy during the season. First thing to point in this sense is that the total amount of absorbed radiation depends much more on the total incident PAR than on the fraction of PAR absorbed (Fig. 5). In other words, the structural effect of drought on green biomass does not have a strong influence on the energy available for the three mentioned processes as the determining factor on the total absorbed energy is sun irradiance. This explains why there are not mainly significant differences in APAR when the effect of drought is happening, during the second part of the growing season (Fig. 4, Panel a).

The seasonal trend of ETR (Fig. 4, Panel b) is highly similar to that of APAR but shows a clear significant decrease upon drought for the susceptible variety. This means that, apart from the structural effect monitored by NDVI, drought has an impact at the physiological level. The reduction of the photosynthetic yield in the second part of the season can be observed in Fig. 7, Panel b, for both tolerant and susceptible varieties. The reduction of PSII quantum yield under drought stress has been observed in *Ricinus communis* L. [69] and is supposed to lead to cyclic electron flow around PSI. Taking into account the lower CO_2 concentration in the mesophyll due to the reduced stomatal conductance caused by water deficit, the reduction in PSII activity can be regarded as a regulated protective mechanism against the generation of reactive oxygen species [70].

Looking at fluorescence emission, Fig. 4, Panel b, shows that, as ETR, SIF emission follows the same trend as that of APAR. This is expected, as more energy reaching the photosystems generates more excited chlorophylls and therefore increased fluorescence emission. The fact that both ETR and SIF are strongly related to the total absorbed energy (Fig. 6, Panels a and b) leads to the correlation between ETR and SIF (Fig. 6, Panel c). This has been observed several times in literature, mainly through GPP measurements through eddy covariance towers, across various crops and regions including crops in China [71], maize in the US Corn Belt [72], wheat [73], and multiple biomes worldwide [24].

Despite the existing correlation between ETR and SIF, fluorescence emission does not reveal significant differences between treatments (Fig. 4, Panel c). This is because although PSII yields decrease upon drought effect (Fig. 7, Panel b), SIF yields remain mostly constant (Fig. 7, Panel c). This lack of correlation between PSII and fluorescence yields (Fig. 8) suggests that in treatments where the energy destined to the photosynthetic process is decreased, the excess energy is not dissipated as fluorescence – otherwise an increase of SIF yield would be observed – but as heat.

According to Porcar-Castell et al. [19], in the absence of NPQ, PSII and SIF yields should be negatively correlated, whereas increasing NPQ tends to be positively correlated. From this point of view, the situation analyzed in this work has to be somewhere in the middle between these two conditions, where correlation between PSII and SIF yield is not significant. By means of continuous ground measurements over maize canopies, Li et al. [74] showed that irradiance-normalized SIF correlates well with light-use efficiency (LUE), but this correlation disappears when canopy structural effects on SIF are taken into account. Also studying maize cultivars, Miao et al. [75] found no correlation between yields during individual growth stages but positive relationships during the season when data is averaged for each stage, reflecting the relevance of the different seasonal-dependent factors. Taking this evidence into account and although much more research on the topic is needed, it seems that it is difficult to find a general relationship between SIF yield and PSII yield or LUE and that we should move towards understanding this relationship in very specific cases.

7 Conclusion

In this work, we have shown how the simultaneous retrieval of both active and passive fluorescence allows the calculation of, first, ETR and SIF and then PSII and SIF yields. Total energy flow through PSII and fluorescence emission is heavily dependent on the total energy absorbed and therefore correlated, but only ETR shows the effect of drought. This is because PSII yield is affected by the inducted stress but SIF yield is not, suggesting that energy not used by PSII is released as heat rather than as fluorescence.

Only through this integrated approach at the interface between leaf level and remote measurements can these conclusions be drawn, given that variable fluorescence techniques are typically applied at the leaf scale and SIF is mostly retrieved from the canopy scale or higher. The new plant monitoring scheme presented in this work provides the necessary link between two very popular but usually unrelated measurement scales. Future research at this interphase scale focusing on different abiotic and biotic stresses will provide more information on the link between photosynthesis and fluorescence that would ultimately allow to estimate photosynthetic rates and yields from aerial vehicles and satellites.

Acknowledgements This work originated from the miniproject "FluoSense4Wheat" funded by HeDWIC supported by the Foundation for Food & Agriculture Research (FFAR) under award number, Grant ID: DFs-19-0000000013. Work of JMR was funded through the country-related cooperation program with Argentina – ALEARG of the German Academic Exchange Service (DAAD) – and through the International Climate Protection Fellowship of the Alexander von Humboldt Foundation. We thank the field team and scientists from Wheat Physiology Group from the International Maize and Wheat Improvement Center, for their support setting up the experiment and supporting the field measurements. We thank the remote sensing technicians at CIMMYT who supported the experiment and data collection: Jorge Montoya Calderón, Javier Padilla Amaya, and Iván Pérez Olivera. We are also grateful to Zbigniew Kolber (Soliense Inc.) for his support with the LIFT device.

References

1. Cooper M, Voss-Fels KP, Messina CD, Tang T, Hammer GL (2021) Tackling G × E × M interactions to close on-farm yield-gaps: creating novel pathways for crop improvement by predicting contributions of genetics and management to crop productivity. Theor Appl Genet 134(6):1625–1644. https://doi.org/10.1007/s00122-021-03812-3
2. Reynolds MP, Braun H-J (2022) Wheat improvement. Springer, Cham. https://doi.org/10.1007/978-3-030-90673-3_1
3. Muruganantham P, Wibowo S, Grandhi S, Samrat NH, Islam N (2022) A systematic literature review on crop yield prediction with deep learning and remote sensing. Remote Sens 14(9):1990
4. Guo J, Zhu Z, Gong P (2024) Global forest resilience change from 2001 to 2022. Int J Remote Sens 45(17):5889–5900
5. Berger K, Machwitz M, Kycko M, Kefauver SC, Van Wittenberghe S, Gerhards M et al (2022) Multi-sensor spectral synergies for crop stress detection and monitoring in the optical domain: a review. Remote Sens Environ 280:113198
6. Agathokleous E, Frei M, Knopf OM et al (2023) Adapting crop production to climate change and air pollution at different scales. Nat Food 4:854–865. https://doi.org/10.1038/s43016-023-00858-y
7. Lagorio MG, Cordon GB, Iriel A, Romero JM, Faivovich J, Taboada C (2020) Biophotonics. Fluorescence and reflectance in living organisms. Sci Rev End World 2(1):18–41
8. Woolley JT (1971) Reflectance and transmittance of light by leaves. Plant Physiol 47(5):656–662
9. Gitelson AA, Zur Y, Chivkunova OB, Merzlyak MN (2002) Assessing carotenoid content in plant leaves with reflectance spectroscopy. Photochem Photobiol 75(3):272–281
10. Zhao F, Huang Y, Guo Y, Reddy KN, Lee MA, Fletcher RS, Thomson SJ (2014) Early detection of crop injury from glyphosate on soybean and cotton using plant leaf hyperspectral data. Remote Sens 6(2):1538–1563
11. Lagorio MG, Cordon GB, Iriel A (2015) Reviewing the relevance of fluorescence in biological systems. Photochem Photobiol Sci 14(9):1538–1559
12. Havaux M (1993) Electron transport system in potato leaves. Plant Sci 94:19–33
13. Ögren E, Öquist G (1985) Effects of drought on photosynthesis, chlorophyll fluorescence and photoinhibition susceptibility in intact willow leaves. Planta 166(3):380–388
14. Ralph PJ, Burchett MD (1998) Photosynthetic response of Halophila ovalis to heavy metal stress. Environ Pollut 103(1):91–101
15. Iriel A, Novo JM, Cordon GB, Lagorio MG (2014) Atrazine and methyl viologen effects on chlorophyll-a fluorescence revisited – implications in photosystems emission and ecotoxicity assessment. Photochem Photobiol 90(1):107–112
16. López Valiño I, Dieguez Gaviola G, Diz VE, González GA, Lagorio MG, Cordon GB (2025) Tracking acetylcholinesterase inhibitor pesticides in the environment using the rapid transient of chlorophyll fluorescence. Photochem Photobiol. https://doi.org/10.1111/php.14080
17. Ač A, Malenovský Z, Olejníčková J, Gallé A, Rascher U, Mohammed G (2015) Meta-analysis assessing potential of steady-state chlorophyll fluorescence for remote sensing detection of plant water, temperature and nitrogen stress. Remote Sens Environ 168:420–436

18. Mohammed GH, Colombo R, Middleton EM, Rascher U, Van Der Tol C, Nedbal L et al (2019) Remote sensing of solar-induced chlorophyll fluorescence (SIF) in vegetation: 50 years of progress. Remote Sens Environ 231:111177
19. Porcar-Castell A, Malenovský Z, Magney T, van Wittenberghe S, Fernández-Marín B, Maignan F, Zhang Y, Maseyk K, Atherton J, Albert LP, Robson TM, Zhao F, Garcia-Plazaola JI, Ensminger I, Rajewicz PA, Grebe S, Tikkanen M, Kellner JR, Ihalainen JA et al (2021) Chlorophyll a fluorescence illuminates a path connecting plant molecular biology to Earth-system science. Nat Plants 7(8):998–1009
20. Cordon GB, Lagorio MG (2007) Optical properties of the adaxial and abaxial faces of leaves. Chlorophyll fluorescence, absorption and scattering coefficients. Photochem Photobiol Sci 6(8): 873–882
21. Cordon G, Iriel A, Cirelli AF, Lagorio MG (2018) Arsenic effects on some photophysical parameters of Cichorium intybus under different radiation and water irrigation regimes. Chemosphere 204:398–404
22. Iriel A, Dundas G, Fernandez Cirelli A, Lagorio MG (2015) Effect of arsenic on reflectance spectra and chlorophyll fluorescence of aquatic plants. Chemosphere 119:697–703
23. van Wittenberghe S, Alonso L, Verrelst J, Hermans I, Delegido J, Veroustraete F et al (2013) Upward and downward solar-induced chlorophyll fluorescence yield indices of four tree species as indicators of traffic pollution in Valencia. Environ Pollut 173:29–37
24. Li X, Xiao J, He B, Altaf Arain M, Beringer J, Desai AR, Emmel C, Hollinger DY, Krasnova A, Mammarella I, Noe SM, Ortiz PS, Rey-Sanchez AC, Rocha AV, Varlagin A (2018) Solar-induced chlorophyll fluorescence is strongly correlated with terrestrial photosynthesis for a wide variety of biomes: first global analysis based on OCO-2 and flux tower observations. Glob Chang Biol 24(9):3990–4008
25. Paruelo JM, Epstein HE, Lauenroth WK, Burke IC (1997) ANPP estimates from NDVI for the central grassland region of the United States. Ecology 78(3):953–958
26. Grigera G, Oesterheld M, Pacín F (2007) Monitoring forage production for farmers' decision making. Agric Syst 94(3):637–648
27. Wieneke S, Balzarolo M, Asard H, Abd Elgawad H, Peñuelas J, Rascher U et al (2022) Fluorescence ratio and photochemical reflectance index as a proxy for photosynthetic quantum efficiency of photosystem II along a phosphorus gradient. Agric For Meteorol 322:109019
28. Drusch M, Moreno J, Del Bello U, Franco R, Goulas Y, Huth A et al (2017) The FLuorescence EXplorer Mission Concept–ESA's Earth Explorer 8. IEEE Trans Geosci Remote Sens 55:1273–1284
29. Bendig J, Siegmann B, Kneer C, Chakhvashvili E, Kraemer J, Choza-Farias S, Rascher U (2023) Imaging spatial heterogeneity of solar-induced chlorophyll fluorescence (SIF) with very high spatial resolution drone imagery. In: Proceedings of the 2023 IEEE International geoscience and remote sensing symposium (IGARSS). IEEE, pp 4586–4589
30. Chang CY, Wen J, Han J, Kira O, LeVonne J, Melkonian J et al (2021) Unpacking the drivers of diurnal dynamics of sun-induced chlorophyll fluorescence (SIF): canopy structure, plant physiology, instrument configuration and retrieval methods. Remote Sens Environ 265:112672
31. Martini D, Sakowska K, Wohlfahrt G, Pacheco-Labrador J, Van der Tol C, Porcar-Castell A et al (2022) Heatwave breaks down the linearity between sun-induced fluorescence and gross primary production. New Phytol 233(6):2415–2428
32. Bendig J, Malenovský Z, Siegmann B, Krämer J, Rascher U (2025) Comparing methods for solar-induced fluorescence efficiency estimation using radiative transfer modelling and airborne diurnal measurements of barley crops. Remote Sens Environ 317:114521
33. Rascher U, Alonso L, Burkart A, Cilia C, Cogliati S, Colombo R, Damm A, Drusch M, Guanter L, Hanus J, Hyvarinen T, Julitta T, Jussila J, Kataja K, Kokkalis P, Kraft S, Kraska T, Matveeva M, Moreno J, Muller O, Panigada C, Pikl M, Pinto F, Prey L, Pude R, Rossini M, Schickling A, Schurr U, Schuttemeyer D, Verrelst J, Zemek F (2015) Sun-induced fluorescence – a new probe of photosynthesis: first maps from the imaging spectrometer HyPlant. Glob Chang Biol 21(12):4673–4684
34. Siegmann B, Cendrero-Mateo MP, Cogliati S, Damm A, Gamon J, Herrera D et al (2021) Downscaling of far-red solar-induced chlorophyll fluorescence of different crops from canopy

to leaf level using a diurnal data set acquired by the airborne imaging spectrometer HyPlant. Remote Sens Environ 264:112609
35. Guanter L, Bacour C, Schneider A, Aben I, van Kempen TA, Maignan F et al (2021) The TROPOSIF global sun-induced fluorescence dataset from the Sentinel-5P TROPOMI mission. Earth Syst Sci Data 13(11):5423–5440
36. Köhler P, Frankenberg C, Magney TS, Guanter L, Joiner J, Landgraf J (2018) Global retrievals of solar-induced chlorophyll fluorescence with TROPOMI: first results and intersensor comparison to OCO-2. Geophys Res Lett 45(19):10–456
37. Magney TS, Bowling DR, Logan BA, Grossmann K, Stutz J, Blanken PD et al (2019) Mechanistic evidence for tracking the seasonality of photosynthesis with solar-induced fluorescence. Proc Natl Acad Sci 116(24):11640–11645
38. Drusch M, Moreno J, Del Bello U, Franco R, Goulas Y, Huth A et al (2016) The fluorescence explorer mission concept – ESA's earth explorer 8. IEEE Trans Geosci Remote Sens 55(3): 1273–1284
39. Machwitz M, Pieruschka R, Berger K, Schlerf M, Aasen H, Fahrner S et al (2021) Bridging the gap between remote sensing and plant phenotyping – challenges and opportunities for the next generation of sustainable agriculture. Front Plant Sci 12:749374
40. Desai AR, Richardson AD, Moffat AM, Kattge J, Hollinger DY, Barr A et al (2008) Cross-site evaluation of eddy covariance GPP and RE decomposition techniques. Agric For Meteorol 148(6–7):821–838
41. Torres R, Romero JM, Lagorio MG (2021) Effects of sub-optimal illumination in plants. Comprehensive chlorophyll fluorescence analysis. J Photochem Photobiol B Biol 218:112182
42. Campbell PK, Huemmrich KF, Middleton EM, Ward LA, Julitta T, Daughtry CS et al (2019) Diurnal and seasonal variations in chlorophyll fluorescence associated with photosynthesis at leaf and canopy scales. Remote Sens 11(5):488
43. Nedbal L, Červený J, Rascher U, Schmidt H (2007) E-photosynthesis: a comprehensive modeling approach to understand chlorophyll fluorescence transients and other complex dynamic features of photosynthesis in fluctuating light. Photosynth Res 93(1):223–234
44. Schurr U, Walter A, Rascher U (2006) Functional dynamics of plant growth and photosynthesis–from steady-state to dynamics–from homogeneity to heterogeneity. Plant Cell Environ 29(3):340–352
45. Krause GH, Jahns P (2003) Pulse amplitude modulated chlorophyll fluorometry and its application in plant science. In: Light-harvesting antennas in photosynthesis. Springer, Dordrecht, pp 373–399
46. Gubsuk W, Banterng P, Vorasoot N, Jogloy S, Theerakulpisut P, Vongcharoen K (2024) Genetic diversity of chlorophyll fluorescence germplasms effects on dry matter of cassava. SABRAO J Breed Genet 56(4):1377
47. Porcar-Castell A, Pfündel E, Korhonen JF, Juurola E (2008) A new monitoring PAM fluorometer (MONI-PAM) to study the short-and long-term acclimation of photosystem II in field conditions. Photosynth Res 96:173–179
48. Kolber ZS, Prášil O, Falkowski PG (1998) Measurements of variable chlorophyll fluorescence using fast repetition rate techniques: defining methodology and experimental protocols. Biochim Biophys Acta Bioenerg 1367(1–3):88–106
49. Osmond B, Chow WS, Wyber R, Zavafer A, Keller B, Pogson BJ, Robinson SA (2017) Relative functional and optical absorption cross-sections of PSII and other photosynthetic parameters monitored in situ, at a distance with a time resolution of a few seconds, using a prototype light induced fluorescence transient (LIFT) device. Funct Plant Biol 44(10):985–1006
50. Keller B, Vass I, Matsubara S, Paul K, Jedmowski C, Pieruschka R, Nedbal L, Rascher U, Muller O (2019) Maximum fluorescence and electron transport kinetics determined by light-induced fluorescence transients (LIFT) for photosynthesis phenotyping. Photosynth Res 140(2): 221–233

51. van der Tol C, Julitta T, Yang P, Sabater N, Reiter I, Tudoroiu M et al (2023) Retrieval of chlorophyll fluorescence from a large distance using oxygen absorption bands. Remote Sens Environ 284:113304
52. Julitta T, Burkart A, Colombo R, Rossini M, Schickling A, Migliavacca M, Cogliati S, Wutzler T, Rascher U (2017) Accurate measurements of fluorescence in the O2A and O2B band using the FloX spectroscopy system – results and prospects. In: Proceedings of the Potsdam GHG Flux Workshop: from photosystems to ecosystems, Potsdam, Germany, 24–26
53. Krämer J, Siegmann B, Castro AO, Muller O, Pude R, Döring T, Rascher U (2025) Downscaling the full-spectrum solar-induced fluorescence emission signal of a mixed crop canopy to the photosystem level using the hybrid approach. Remote Sens Environ 324:113000
54. Cendrero-Mateo MP, Wieneke S, Damm A, Alonso L, Pinto F, Moreno J et al (2019) Sun-induced chlorophyll fluorescence III: benchmarking retrieval methods and sensor characteristics for proximal sensing. Remote Sens 11(8):962
55. Sippel S, Reichstein M, Ma X, Mahecha MD, Lange H, Flach M, Frank D (2018) Drought, heat, and the carbon cycle: a review. Curr Clim Chang Rep 4:266–286
56. Zendonadi dos Santos N, Piepho HP, Condorelli GE, Licieri Groli E, Newcomb M, Ward R et al (2021) High-throughput field phenotyping reveals genetic variation in photosynthetic traits in durum wheat under drought. Plant Cell Environ 44(9):2858–2878
57. Damm A, Cogliati S, Colombo R, Fritsche L, Genangeli A, Genesio L et al (2022) Response times of remote sensing measured sun-induced chlorophyll fluorescence, surface temperature and vegetation indices to evolving soil water limitation in a crop canopy. Remote Sens Environ 273:112957
58. Zendonadi dos Santos N (2022) High-throughput phenotyping of photosynthesis traits in durum wheat under drought stress using light-induced fluorescence transients. Doctoral dissertation. Rheinische Friedrich-Wilhelms-Universitaet, Bonn
59. Tucker CJ (1979) Red and photographic infrared linear combinations for monitoring vegetation. Remote Sens Environ 8:127–150
60. Alonso L, Gómez-Chova L, Vila-Francés J, Amorós-López J, Guanter L, Calpe J, Moreno J (2008) Improved Fraunhofer line discrimination method for vegetation fluorescence quantification. IEEE Geosci Remote Sens Lett 5(4):620–624
61. Moreau L, Li Z (1996) A new approach for remote sensing of canopy absorbed photosynthetically active radiation. II: proportion of canopy absorption. Remote Sens Environ 55(3): 192–204
62. Zeng Y, Badgley G, Dechant B, Ryu Y, Chen M, Berry JA (2019) A practical approach for estimating the escape ratio of near-infrared solar-induced chlorophyll fluorescence. Remote Sens Environ 232:111209
63. Mattera J, Romero JM, Irisarri JGN, Grimoldi AA, Cordon GB (2025) Deciphering the link between SIF, ANPP and RUE dynamics in perennial forage crop mixtures. Agric For Meteorol 369:110569
64. White JW, Conley MM (2013) A flexible, low-cost cart for proximal sensing. Crop Sci 53(4): 1646–1649
65. Huete A, Didan K, Miura T, Rodriguez EP, Gao X, Ferreira LG (2002) Overview of the radiometric and biophysical performance of the MODIS vegetation indices. Remote Sens Environ 83(1–2):195–213
66. Liu L, Yang X, Zhou H, Liu S, Zhou L, Li X et al (2018) Evaluating the utility of solar-induced chlorophyll fluorescence for drought monitoring by comparison with NDVI derived from wheat canopy. Sci Total Environ 625:1208–1217
67. Zhang J, Xiao J, Tong X, Zhang J, Meng P, Li J et al (2022) NIRv and SIF better estimate phenology than NDVI and EVI: effects of spring and autumn phenology on ecosystem production of planted forests. Agric For Meteorol 315:108819
68. Thénot F, Méthy M, Winkel T (2002) The photochemical reflectance index (PRI) as a water-stress index. Int J Remote Sens 23(23):5135–5139

69. Lima Neto MC, Silveira JAG, Cerqueira JV, Cunha JR (2017) Regulation of the photosynthetic electron transport and specific photoprotective mechanisms in Ricinus communis under drought and recovery. Acta Physiol Plant 39(8):1–12
70. Zahra N, Hafeez MB, Kausar A, Al Zeidi M, Asekova S, Siddique KH, Farooq M (2023) Plant photosynthetic responses under drought stress: effects and management. J Agron Crop Sci 209(5):651–672
71. Cui T, Sun R, Qiao C (2017) Analyzing the relationship between solar-induced chlorophyll fluorescence and gross primary production using remotely sensed data and model simulation. Int J Earth Environ Sci 2(129):10
72. Wagle P, Zhang Y, Jin C, Xiao X (2016) Comparison of solar-induced chlorophyll fluorescence, light-use efficiency, and process-based GPP models in maize. Ecol Appl 26(4): 1211–1222. https://doi.org/10.1890/15-1434
73. Liu L, Liu X, Hu J, Guan L (2017) Assessing the wavelength-dependent ability of solar-induced chlorophyll fluorescence to estimate the GPP of winter wheat at the canopy level. Int J Remote Sens 38(15):4396–4417
74. Li Z, Zhang Q, Li J, Yang X, Wu Y, Zhang Z, Wang S, Wang H, Zhang Y (2020) Solar-induced chlorophyll fluorescence and its link to canopy photosynthesis in maize from continuous ground measurements. Remote Sens Environ 236. https://doi.org/10.1016/j.rse.2019.111420
75. Miao G, Guan K, Suyker AE, Yang X, Arkebauer TJ, Walter-Shea EA, Kimm H, Hmimina GY, Gamon JA, Franz TE, Frankenberg C, Berry JA, Wu G (2020) Varying contributions of drivers to the relationship between canopy photosynthesis and far-red sun-induced fluorescence for two maize sites at different temporal scales. J Geophys Res Biogeosci 125(2). https://doi.org/10.1029/2019JG005051

From Space to Cells: Physical Models for Light Re-absorption and Scattering

María Gabriela Lagorio, Gabriela Beatriz Cordon, Juan Manuel Romero, and Ivana López Valiño

Contents

The original version of the chapter has been revised. A correction to this chapter can be found at https://doi.org/10.1007/17361_2025_13

M. G. Lagorio (✉)
Facultad de Ciencias Exactas y Naturales, Dpto. de Química Inorgánica, Analítica y Química Física, Universidad de Buenos Aires, Buenos Aires, Argentina

CONICET, Instituto de Química Física de los Materiales, Medio Ambiente y Energía (INQUIMAE), Facultad de Ciencias Exactas y Naturales, Universidad de Buenos Aires, Buenos Aires, Argentina
e-mail: mgl@qi.fcen.uba.ar

G. B. Cordon
Facultad de Agronomía, Área de Educación Agropecuaria y Ambiental, Universidad de Buenos Aires, Buenos Aires, Argentina

CONICET, Instituto de Investigaciones Fisiológicas y Ecológicas vinculadas a la Agricultura (IFEVA), Facultad de Agronomía, Universidad de Buenos Aires, Buenos Aires, Argentina
e-mail: gcordon@agro.uba.ar

J. M. Romero
Facultad de Ciencias Exactas y Naturales, Dpto. de Química Inorgánica, Analítica y Química Física, Universidad de Buenos Aires, Buenos Aires, Argentina

CONICET, Instituto de Investigaciones Fisiológicas y Ecológicas vinculadas a la Agricultura (IFEVA), Facultad de Agronomía, Universidad de Buenos Aires, Buenos Aires, Argentina

I. López Valiño
Facultad de Ciencias Exactas y Naturales, Dpto. de Química Inorgánica, Analítica y Química Física, Universidad de Buenos Aires, Buenos Aires, Argentina

Facultad de Agronomía, Área de Educación Agropecuaria y Ambiental, Universidad de Buenos Aires, Buenos Aires, Argentina
e-mail: ivalino@qi.fcen.uba.ar

M. G. Lagorio (ed.), Comprehensive Series in Photochemical and Photobiological Sciences, Fluorescence of Living Plants 1, https://doi.org/10.1007/17361_2024_2
Published online: 25 March 2025

Abstract The analysis of chlorophyll-a fluorescence in plants is a valuable tool for gaining insights into plant physiology, photosynthetic performance, and potential alterations at the level of photosystems I and II. One method of recording fluorescence involves measuring the emitted intensity as a function of wavelength. These spectra are typically obtained under steady-state conditions, where the spectrum remains constant over time, unaffected by variable fluorescence induction. At room temperature, the fluorescence spectrum of chlorophyll-a in plants is characterised by two bands: one in the red region due to the emission of photosystem II and one in the far-red region linked to emissions from both photosystems. The ratio of the red to far-red peaks has been correlated with the presence of biotic and abiotic factors in the environment. However, fluorescence originating at the chloroplast level undergoes light re-absorption and re-emission processes within the plant leaves. Shorter wavelength photons are more likely to be absorbed by the tissue than longer wavelength photons, leading to a distortion in the signal recorded outside the leaves. This phenomenon also occurs at the canopy level. The plant physiology relates to the original spectrum emitted by the chloroplast, not to the distorted spectrum observed externally. Therefore, it is essential to correct the experimentally observed spectra for these re-absorption and re-emission processes. This chapter describes physical models for correcting both the spectral distribution of fluorescence and the fluorescence quantum yields.

Keywords Chlorophyll fluorescence, Inner filter, Light re-absorption, Modelling chlorophyll fluorescence

1 Introduction

The fluorescence spectra of chlorophyll-a from plant leaves (plotted as fluorescence intensity as a function of wavelength) are characterised by two bands: one in the red, attributed to Photosystem II, and a second in the far red, arising from the emission of both Photosystems II and I (Fig. 1) [1, 2]. Due to the overlap between the absorption and emission spectra of a leaf, fluorescence re-absorption takes place. In fact, on their way out of the leaf, fluorescence photons are partially absorbed and re-emitted within the plant tissue according to their wavelength, causing spectral distortion and

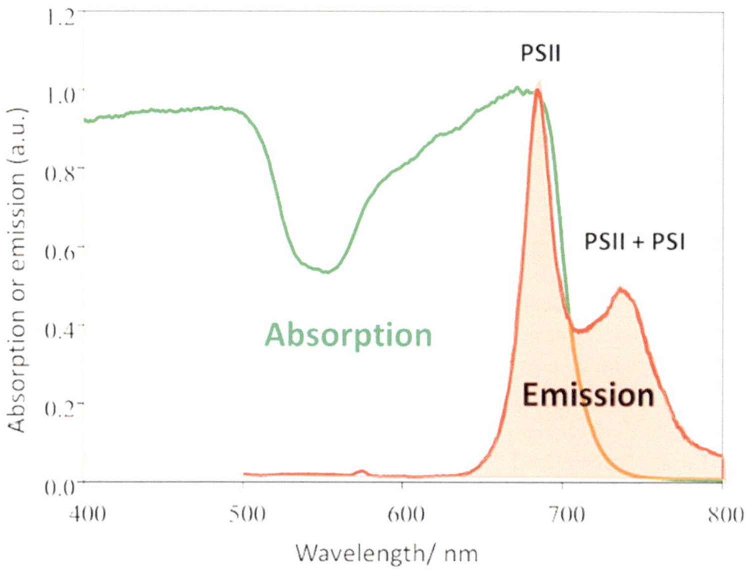

Fig. 1 Typical absorption spectrum (green line) and chlorophyll-a fluorescence emission (red line) of a plant leaf. The absorption spectrum was calculated as the remission function: $F(R) = (1 - R)^2/(2R)$ with R: reflectance of a thick layer of leaves [7]. The overlap between both spectra mainly affects the red emission band

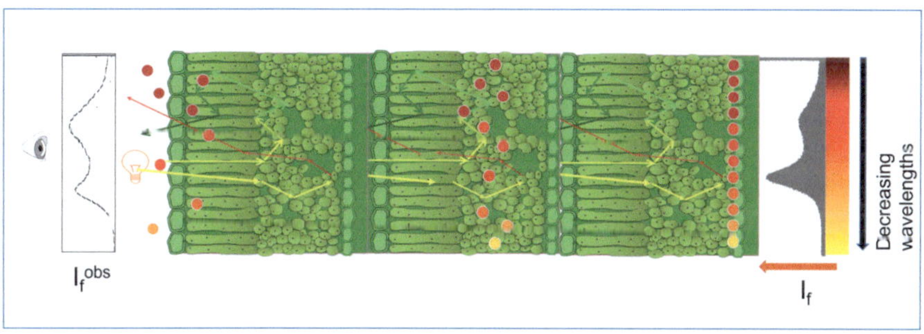

Fig. 2 Illustration of the fluorescence re-absorption process in plant tissue: Fluorescence photons with shorter wavelengths (higher energy) are more strongly absorbed by the plant tissue than photons with longer wavelengths. In this illustration, luminescence originates from the right side of the image (at the chloroplasts) and photons travel towards the left side, where an external detector records the experimental signal (I_f^{obs})

attenuation of the light that reaches an external detector [1, 3–5]. The distortion of the fluorescent signal emitted by chloroplasts is notably significant due to the high pigment concentration and large scattering in the plant tissue (Fig. 2). These processes affect the peak ratio (known as fluorescence ratio $F_{red}/F_{far\text{-}red}$), with the effect being greater the higher the chlorophyll concentration. The decrease in the red/

far-red quotient with increasing chlorophyll concentration has prompted researchers to explore the relationship between chlorophyll content and the fluorescence ratio [6]. This ratio has been used as a non-destructive optical method to estimate the surface concentration of these photosynthetic pigments in leaves [5].

One of the challenges in studying the stationary fluorescence of chlorophyll-a is to recover the fluorescence emitted by chloroplasts before it is attenuated by the internal filtering effect (I_f) from the experimentally observed spectrum from leaves (I_f^{obs}). The recovery of this corrected signal should be achieved in both spectral distribution and intensity (the total number of photons). Concerning the fluorescence ratio, while the "experimental" or "observed" quotient is inversely related to chlorophyll content, the fluorescence ratio corrected for re-absorption and re-emission of light is independent of pigment concentration and directly related to photosystem stoichiometry, plant physiology, and/or the presence of biotic or abiotic stresses, especially those that block electron transfer from PSII to PSI [8–13]. Moreover, when the fluorescence signal is remotely registered, emission from leaves can suffer additional re-absorption and scattering within the canopy [14]. It is therefore clear that a correction for re-absorption and re-emission processes is relevant for the correct interpretation of both fluorescence intensity and fluorescence ratio. The main objective of this chapter is the presentation and discussion of the physical models related to this correction, with emphasis on the two aspects under consideration: fluorescence attenuation and spectral distortion.

2 Addressing Fluorescence Attenuation: Correcting Quantum Yields and Fluorescence Intensity

Let us consider our system as a collection of fluorescent units that emit ϕ fluorescence photons per absorbed photon. Some of these emitted photons may be re-absorbed within the system before exiting [15–17]. Some of the reabsorbed energy will be re-emitted as fluorescence and so on. Consequently, in addition to the spectral distortion caused by light re-absorption, there is also a decrease in the observed fluorescence quantum yield ϕ_{obs} compared to ϕ. If I_a represents the number of photons absorbed by the system, then I_a times the fluorescence quantum yield ($I_a \cdot \phi$) states for the number of fluorescent photons emitted. These initially emitted photons have a chance of being re-absorbed with a probability P_1. Consequently, $I_a \cdot \phi \cdot P_1$ photons will be re-absorbed, while $I_a \cdot \phi \cdot (1 - P_1)$ will exit the system. This re-absorbed fraction will generate $I_a \cdot \phi^2 \cdot P_1$ photons, of which a portion $I_a \cdot \phi^2 \cdot P_1 \cdot P_2$ will be re-absorbed, leaving $I_a \cdot \phi^2 \cdot P_1 .(1 - P_2)$ to escape. This process continues for n steps, resulting in a net number of escaping photons calculated as the sum: $I_a \cdot \phi \cdot (1 - P_1) + I_a \cdot \phi^2 \cdot P_1 \cdot (1 - P_2) + I_a \cdot \phi^3 \cdot P_1 P_2 \cdot (1 - P_3) + \ldots I_a \cdot \phi^n \cdot P_1 P_2 \ldots P_{n-1} \cdot (1 - P_n)$. The observed fluorescence quantum yield ϕ_{obs} is then determined by the ratio of the sum of all emerging photons (as n approaches infinity, $n \to \infty$) to I_a [17–19] (Eq. (1)).

$$\phi_{obs} = \frac{I_a.\varphi.(1-P_1)+I_a.\varphi^2.P_1.(1-P_2)+I_a.\varphi^3.P_1P_2.(1-P_3)+\ldots I_a.\varphi^n.P_1P_2\ldots P_{n-1}.(1-P_n)}{I_a},$$

$$\phi_{obs} = \phi\,(1-P_1) + \sum_{n=2}^{\infty} \phi^n (1-P_n) \prod_{i=1}^{n-1} P_i \tag{1}$$

Suppose that $P_1 = P_2 = P_3 = \ldots = P_i = P$ meaning the probabilities of re-absorption remain constant regardless of the number of re-absorption and re-emission steps. After some manipulation of Eq. (1), utilizing the known result for the sum of an infinite geometric sequence,[1] Eq. (2) was derived to give the fluorescence quantum yield without the influence of light re-absorption artefacts:

$$\phi = \frac{\phi_{obs}}{1 - P(1-\phi_{obs})} \tag{2}$$

To calculate the factor P, we consider that the total number of detected photons emitted by the sample can be calculated by Eq. (3)

$$J = I_a.\phi\,(1-P) \tag{3}$$

If we denote as $\gamma(\lambda, \lambda_0)$, the escape probability of a photon emitted at λ when the sample is excited at λ_0, the number of emitted photons at each emission wavelength can be estimated as

$$j(\lambda) = I_a.\phi\; g(\lambda).\gamma(\lambda, \lambda_0) \tag{4}$$

where $g(\lambda)$ is the spectral distribution of fluorescence free from light re-absorption processes.

Integration of Eq. (4) for the whole range of emission wavelengths gives Eq. (5)

$$J = I_a.\phi \int_\lambda g(\lambda).\gamma(\lambda, \lambda_0)d\lambda \tag{5}$$

Comparing Eqs. (3) and (5), an expression for the P factor is obtained [18, 19].

$$P = 1 - \int_\lambda g(\lambda).\gamma(\lambda, \lambda_0)d\lambda, \quad \text{with} \int_\lambda g(\lambda) = 1 \tag{6}$$

It is important to notice that Eq. (2) is general and does not depend on the system under study. The only assumption involved is that the probabilities of re-absorption remain constant regardless of the number of light re-absorption and re-emission steps.

---

[1] $1 + \phi P + \phi^2 P^2 + \phi^3 P^3 + \cdots + \phi^n P^n = 1/(1 - \phi P)$.

Table 1 Examples of photosynthetic systems and their respective fluorescent units for $g(\lambda)$ estimation

System	Fluorescent unit for $g(\lambda)$ estimation
Leaf 	Chloroplast 
Canopy 	Leaf 

From the ϕ value, considering it is defined as the number of emitted photons per absorbed photon, the corrected number of fluorescence photons that should exit the system in the absence of re-absorption processes can be calculated as $I_a \cdot \phi$ [20].

Consequently, the correction for fluorescence attenuation by using Eq. (2) requires:

(i) Determining the experimental fluorescence quantum yield ϕ_{obs}, i.e., the ratio between the total experimentally recorded fluorescence intensity and the amount of absorbed light.
(ii) Calculating the factor P, which represents the probability that a photon emitted by fluorescence will be re-absorbed. Estimation of this factor could be achieved by recording $g(\lambda)$ and calculating $\gamma\ (\lambda, \lambda_0)$ through physical models. In the following sections, physical approaches for estimating $\gamma(\lambda, \lambda_0)$, applicable to diverse systems (leaves or canopies), will be presented.

It should be noted that Eq. (3) is valid for any type of fluorescent material and it may be applied to opaque, translucent, or transparent samples. However, the function $\gamma(\lambda, \lambda_0)$ depends on the type of system, the measurement geometry, and the absorption of the fluorophore at the excitation and emission wavelengths.

When analysing steady-state chlorophyll fluorescence at different levels, distortions in the emitted spectra are inevitable. A leaf can be considered as a system composed of chloroplasts acting as fluorescent units, whereas a canopy can be visualised as a system made up of leaves as fluorescence sources. In this way, to retrieve the chloroplast-level fluorescence from the leaf's emission, the normalised fluorescence spectrum of a thin layer of isolated chloroplasts can serve as an estimator of $g(\lambda)$ [3]. Likewise, to infer the leaf-level spectrum from the canopy-level data, the normalised fluorescence spectrum of a leaf can be utilized to estimate the distribution function $g(\lambda)$[2] [21, 22]. In Table 1, examples of photosynthetic systems and their respective fluorescent units are presented.

If desired, calculation of the apparent fluorescence quantum yield may also be performed as $I_a.\phi/I_0$, where I_0 is the number of incident photons.

[2]In other practical applications that do not involve plant leaves, but rather different types of fluorescent materials, $g(\lambda)$ can be obtained by recording the spectrum of a very dilute solution (for liquid samples) or by registering the spectrum of a very thin layer of the sample (for solids) [18].

3 Addressing Spectral Distortion: Correcting the Fluorescence Ratio

3.1 Leaf to Chloroplasts

In 1993, Agati et al. developed a theoretical model to correct the experimental fluorescence spectra of leaves [4]. The main feature of this model is assuming that the processes of light attenuation by absorption and/or re-absorption are monoexponential.

In order to apply the correction for the re-absorption and scattering of light within this model, the experimental fluorescence spectrum is divided by the function $\gamma(\lambda, \lambda_0)$, which is calculated from the experimental reflectance and transmittance of a leaf at the excitation and emission wavelengths.

$$I_f^c(\lambda) = \frac{I_f^e(\lambda)}{\gamma(\lambda, \lambda_0)} \tag{7}$$

$$\gamma(\lambda, \lambda_0) = \left[\frac{\ln 1/(r_{\lambda_0}+t_{\lambda_0}) + \ln 1/(r_\lambda+t_\lambda)}{\ln 1/(r_{\lambda_0}+t_{\lambda_0})} \frac{1 - r_{\lambda_0} - t_{\lambda_0}}{1 - (r_{\lambda_0} + t_{\lambda_0}).(r_\lambda + t_\lambda)} \right]^{-1} \tag{8}$$

where λ and λ_0 state for the emission and excitation wavelengths, respectively. The parameters r and t represent reflectance and transmittance of a single leaf, respectively (Fig. 4).

This model is applicable when fluorescence detection and excitation are performed on the same side and the following assumptions are involved in its derivation:

(i) An infinitesimal layer within the leaf is considered.
(ii) The excitation beam which reaches the infinitesimal layer is attenuated monoexponentially within the leaf due to the light absorption by the pigments.
(iii) Part of the absorbed light at the infinitesimal layer is emitted as fluorescence.
(iv) The photons emitted by the infinitesimal layer are attenuated monoexponentially by the process of light re-absorption.
(v) Fluorescence emission within the leaf is isotropic.
(vi) The leaf area is large compared to its thickness and lateral light losses are negligible.
(vii) Light scattering is included in the approach by defining an effective path length.
(viii) The fraction of non-absorbed radiation is calculated as the sum of leaf reflectance and transmittance.
(ix) Monochromatic excitation.

To validate the model, the authors recorded fluorescence spectra from both the wild type and the *aurea* mutant tomato leaves, which have different chlorophyll content and chloroplasts ultrastructure. The authors hypothesized that the observed spectral differences were solely due to variations in light re-absorption processes in each type. The agreement between corrected spectra demonstrated the model validity. A detailed derivation of the correction factor may be found in the original article [4].

In 1998, Lagorio et al. developed a two-flux model to correct light re-absorption processes in systems where a fluorophore was immobilised on inert supports [19]. Later, Ramos and Lagorio applied this model to plant leaves [3]. This approach is based on a physical description of the light interaction with the leaf based on the Kubelka-Munk theory [7].

As with the previous model, this one is also applicable when excitation and fluorescence detection occur on the same side. The assumptions involved in its derivation are listed below:

(i) The system is composed of a thick layer of leaves illuminated normally to the surface.
(ii) It behaves as a homogeneous ideal scattering and absorbing medium.
(iii) No light is transmitted from the group of leaves (opaque system).
(iv) Two light fluxes are considered.
(v) Fluorescence is generated in each volume element within the leaf. The emitted light is decomposed into two-photon flows with opposite directions.
(vi) The two-photon flows for the emitted light have the same magnitude.
(vii) Monochromatic excitation.

The correction factor predicted by this approach is given by Eq. (9). For a detailed derivation of the correction factor, see [19] and the Appendix of this chapter.

$$\gamma(\lambda,\lambda_0)=\frac{1}{1+\sqrt{\frac{F(R_\lambda)}{F(R_\lambda)+2}}}\,\frac{1}{1+\sqrt{\frac{F(R_\lambda)(F(R_\lambda)+2)}{F\left(R_{\lambda_0}\right)\left(F\left(R_{\lambda_0}\right)+2\right)}}} \tag{9}$$

where $F(R_\lambda)$ is the remission function at wavelength (λ), which is calculated from the diffuse reflectance for a thick layer of leaves (R_λ) [7, 23].

$$F(R_\lambda)=\frac{(1-R_\lambda)^2}{2R_\lambda} \tag{10}$$

Alternatively, $F(R)$, which is the ratio between the absorption coefficient (k) and the scattering coefficient (s) (Eq. (11)), can be calculated from the reflectance (r) and transmittance (t) of single leaves by using Eqs. (12), (13), (14), (15), and (16) derived from the theory of pile of plates [23–28]. The coefficient k is directly related to the chromophore concentration C by the equation $k = 2\varepsilon C$, where ε is the molar Napierian absorption coefficient, and C is the chromophore concentration expressed in mol per unit volume. Similarly, the coefficient s is given by $Ss = 2\sigma$, where σ states for the fraction of light scattered per unit path length in the sample.

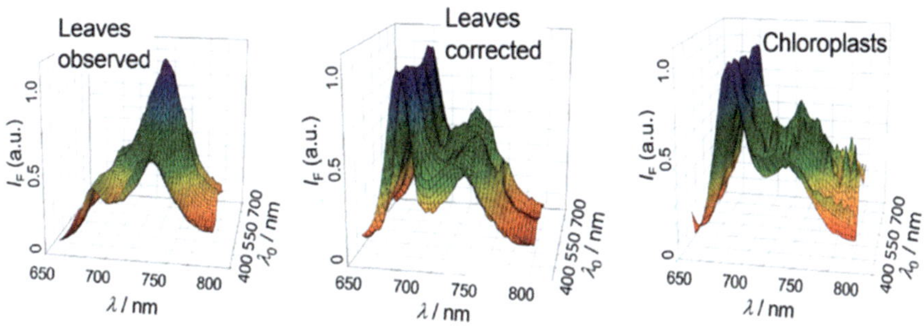

Fig. 3 Fluorescence emission observed for *Ficus benjamina* leaves at different wavelengths. Left: experimental spectra from intact leaves, centre: spectra corrected for light re-absorption processes by Eqs. (7) and (9), right: spectra for a thin layer of chloroplasts deposited on a glass plate. All the determinations were performed in front-face geometry in a steady-state fluorometer. The photon flux used for the excitation beam was low enough as not to induce the variable fluorescence of chlorophyll-a. For more details, see [1, 3]. Adapted from [3] with permission from the European Society for Photobiology, the European Photochemistry Association, and The Royal Society of Chemistry

$$F(R_\lambda) = \frac{k}{s} \tag{11}$$

$$k = \frac{a-1}{a+1}\log b \tag{12}$$

$$s = \frac{2a}{a^2-1}\log b \tag{13}$$

$$a = \frac{1+r^2-t^2+\Delta}{2r} \tag{14}$$

$$b = \frac{1+rt^2-t^2+\Delta}{2t} \tag{15}$$

$$\Delta^2 = (1+r+t)(1+r-t)(1-r+t)(1-r-t) \tag{16}$$

Once the correction factor is known (Eq. (9)), the corrected spectral distribution of fluorescence can be calculated by using Eq. (7).

To test the validity of this model in plants, the authors compared the shape of the emission spectra of intact leaves corrected for light re-absorption with the fluorescence spectral distribution of thin layers of chloroplasts, upon excitation at different wavelengths [3]. Figure 3 shows these results, where it should be noted that the spectra for chloroplasts and leaves corrected for light re-absorption are normalized in the red peak. Therefore, the comparison should focus only on the spectral shape and not on the absolute intensity values.

In summary, by applying any of the methods described in this section, it is possible to retrieve the **spectral shape of the fluorescence at chloroplast level from the observed spectral distribution at leaf level.** Furthermore, assuming $g(\lambda)$ is experimentally available, it is possible to **correct for photon flux attenuation** due

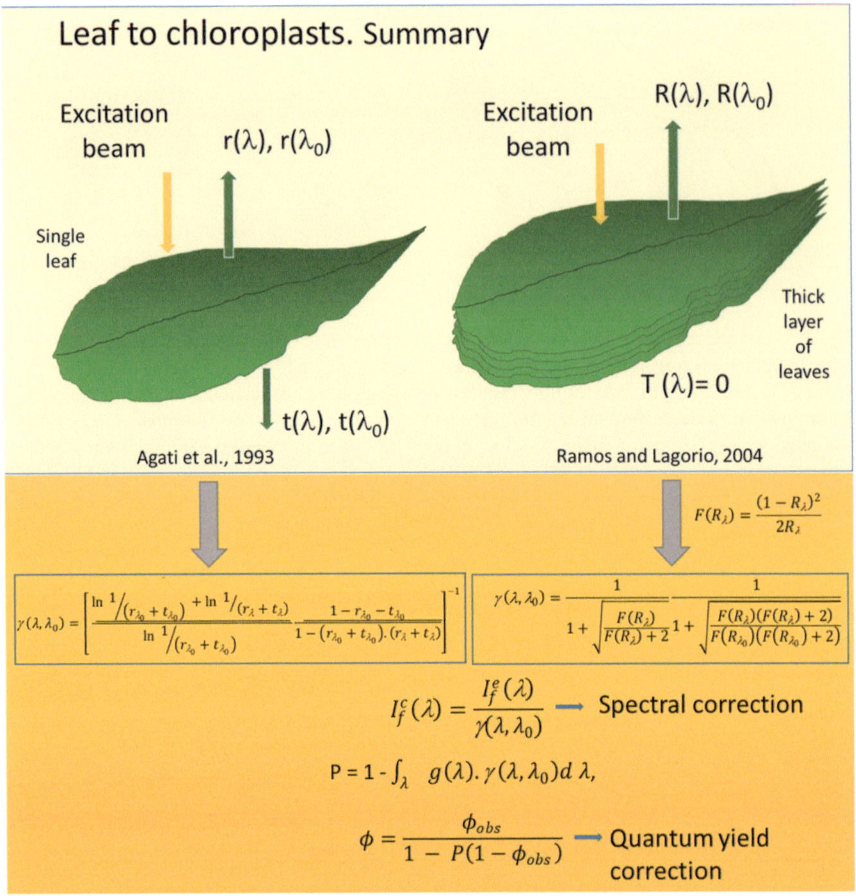

Fig. 4 The upper part of the figure shows the experimental variables to be measured in order to perform the correction for light re-absorption processes in leaves, either in the model described by [4] or by [3]. While in the first model, reflectance and transmittance of single leaves ($r(\lambda)$, $r(\lambda_0)$, $t(\lambda)$, $t(\lambda_0)$) are needed, in the second model the reflectance of thick (non-transmitting) layers ($R(\lambda)$, $R(\lambda_0)$) of leaves are required to satisfy the model assumptions. The lower part of the figure summarises the equations needed for the calculation of the correction function (γ (λ, λ_0)) according to each model. The equations for corrections of both the spectral distribution and the fluorescence quantum yield are also included. To obtain experimentally the distribution $g(\lambda)$, the fluorescence of a thin layer of chloroplasts on a glass plate can be recorded

to light re-absorption and scattering as explained in sect. 2. To proceed with this last step, the corrected quantum yield is calculated and then multiplied by the total absorbed photon flux by the plant. The result of this calculation represents the total number of fluorescence photons emitted by chloroplasts before experiencing scattering and filter effects within the leaves. Thus, both the spectral distribution and the fluorescence intensity may be corrected (Fig. 4).

The topic of light re-absorption in leaves has been widely discussed in the literature from a qualitative point of view [29, 30]. However, the quantification of this phenomenon has been subject of controversy and debates in the literature. A discussion on different correction models at leaf level can be found in [1].

3.2 Canopy to Leaf

In 2018, Romero et al. [22] developed a model to retrieve the fluorescence spectrum of chlorophyll-a and the fluorescence quantum yield at leaf level, from data observed for the whole canopy. The model was revised and improved in 2020 by the same authors [21]. This model was inspired in the approach derived by Agati et al. [4] for single leaves.

For this model, the correction function is given by Eq. (17)

$$
\begin{aligned}
\gamma(\lambda,\lambda_0) &= \frac{1}{\chi(\lambda,\lambda_0)} \\
&= \left[\frac{1}{1+R_{\mathrm{b}}(\lambda_0).T_{\mathrm{c}}(\lambda_0)}\right]\left[\frac{\ln\xi(\lambda_0)}{\ln\xi(\lambda_0)+\ln\xi(\lambda)}\right]\left[\frac{1-\xi(\lambda_0)\,\xi(\lambda)}{1-\xi(\lambda_0)}\right] \\
&\quad+\left[\frac{R_{\mathrm{b}}(\lambda_0).T_{\mathrm{c}}(\lambda_0)}{1+R_{\mathrm{b}}(\lambda_0).T_{\mathrm{c}}(\lambda_0)}\right]\left[\frac{\ln\xi(\lambda_0)}{\ln\xi(\lambda)-\ln\xi(\lambda_0)}\right]\left[\frac{\xi(\lambda_0)-\xi(\lambda)}{1-\xi(\lambda_0)}\right]
\end{aligned}
\tag{17}
$$

where $R_{\mathrm{b}}(\lambda_0)$ is the background (soil) reflectance at the excitation wavelength, $T_{\mathrm{c}}(\lambda_0)$ is the transmittance through the canopy at the excitation wavelength, and $\xi(\lambda)$ is the fraction of light non-absorbed by the canopy which can be calculated through Eq. (18).

$$
\xi(\lambda) = R + (1-C_{\mathrm{f}})(1-R_{\mathrm{b}}) + C_{\mathrm{f}}\cdot T_{\mathrm{c}}\cdot(1-R_{\mathrm{b}})\cdot\left(\frac{1}{1-r}\right) \tag{18}
$$

In Eq. (18), R stands for the total reflectance of the system (canopy + soil), C_{f} is the fraction of soil covered by vegetation, and r is the product between R_{b} and the reflectance of the vegetation measured from the soil (R'_{c}).

The correction function $\gamma(\lambda,\lambda_0)$ is then used in Eq. (7) to obtain the corrected spectral distribution of fluorescence.

A detailed deduction of these equations appears in reference [21] and in the Appendix.

In practice, $R_{\mathrm{b}}(\lambda_0)$ and $T_{\mathrm{c}}(\lambda_0)$ display low values[3] and the second term in Eq. (17) may be neglected in most of the cases, resulting in an approximate expression for the correction function given by Eq. (19):

[3] Soil is usually dark and has low values of reflectance. Compact vegetation transmits a poor quantity of photons and displays low transmittance.

$$\gamma(\lambda,\lambda_0) \sim \frac{1}{\chi(\lambda,\lambda_0)} = \left[\frac{1}{1+R_b(\lambda_0).T_c(\lambda_0)}\right]\left[\frac{\ln\xi(\lambda_0)}{\ln\xi(\lambda_0)+\ln\xi(\lambda)}\right]\left[\frac{1-\xi(\lambda_0)\,\xi(\lambda)}{1-\xi(\lambda_0)}\right] \tag{19}$$

In the model, it is assumed that:

(i) The system under study is composed of the canopy and the soil.
(ii) The whole system does not transmit light.
(iii) Light is monoexponentially attenuated within the canopy due to both absorption and re-absorption processes.
(iv) The vertical structure of the canopy is homogeneous.
(v) There is re-absorption of the fluorescence emitted from the leaves by the canopy, so that it is attenuated before emerging from the vegetation cover.
(vi) The scattering of light within the canopy is considered by defining an effective path length.
(vii) Multi-scattering between the soil and the canopy is explicitly considered.
(viii) The soil fractions not covered by plants directly reflect the incident light. The amount of reflected light by these areas is given by the soil reflectance at the excitation wavelength.
(ix) Monochromatic excitation.

The physical approach was validated by comparing the corrected spectra of a canopy with the spectral distribution of fluorescence at leaf level.

In summary, by applying this model, it is possible to retrieve the **spectral shape of the fluorescence at leaf level from the observed spectral distribution at canopy level.** Additionally, assuming $g(\lambda)$ is experimentally available, it is possible to **correct for photon flux attenuation** due to light re-absorption and scattering as explained in sect. 2. Thus, the corrected quantum yield can be calculated and then multiplied by the total absorbed photon flux by the vegetation cover, to yield the total number of fluorescence photons emitted by the leaves before experiencing scattering and filter effects within the canopy (Fig. 5).

In conclusion, by the consecutive application of the canopy-to-leaf and leaf-to-chloroplast models, it is possible to obtain information at the microscopic level from data recorded at distances about a million times greater (Fig. 6).

4 About the Calculation of the Observed Fluorescence Quantum Yield (ϕ_{obs})

To complete the description of the methodology used for inner filter and scattering corrections, let us examine how the observed fluorescence quantum yield ϕ_{obs} can be experimentally calculated at both leaf and canopy levels.

In general, for chlorophyll-a emission, which is observed between 600 and 800 nm, ϕ_{obs} can be calculated by Eq. (20):

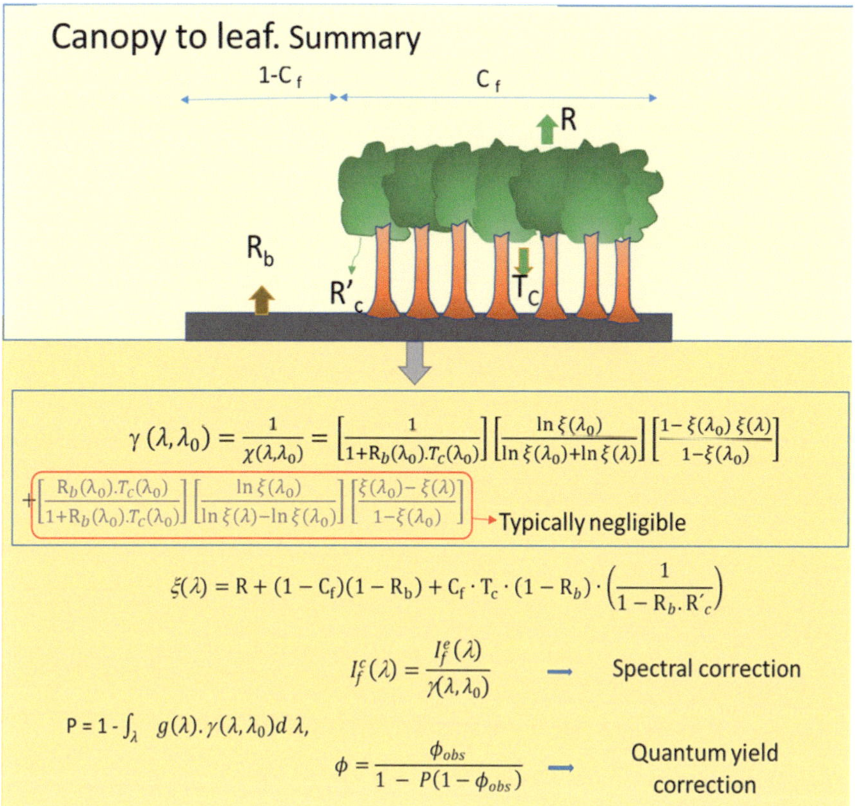

Fig. 5 The upper part of the figure shows the experimental variables to be measured in order to perform the correction for light re-absorption processes in canopies following [21]. The lower part of the figure summarises the equations needed for the calculation of the correction function ($\gamma(\lambda, \lambda_0)$) according to this model. The equations for corrections of both the spectral distribution and the fluorescence quantum yield are also included. To obtain experimentally the distribution $g(\lambda)$, the fluorescence of a single leaf can be recorded

$$\phi_{\text{obs}} = \frac{\int_{600}^{800} I_{\text{F}}(\lambda)^{\text{exp}}.d\lambda}{\int_{400}^{800} I_{\text{a}}(\lambda_0).d\lambda_0} \tag{20}$$

where $I_{\text{F}}(\lambda)^{\text{exp}}$ is the experimental fluoresce spectra and $I_{\text{a}}(\lambda_0)$ is the absorbed photon flux at the excitation wavelength.

The numerator (representing the total number of emitted photons) is integrated over the chlorophyll-a emission range, and the denominator (representing the total number of absorbed photons) is integrated over the interval of wavelengths involved in the excitation. In the case of monochromatic excitation, the denominator reduces simply to $I_{\text{a}}(\lambda_0)$.

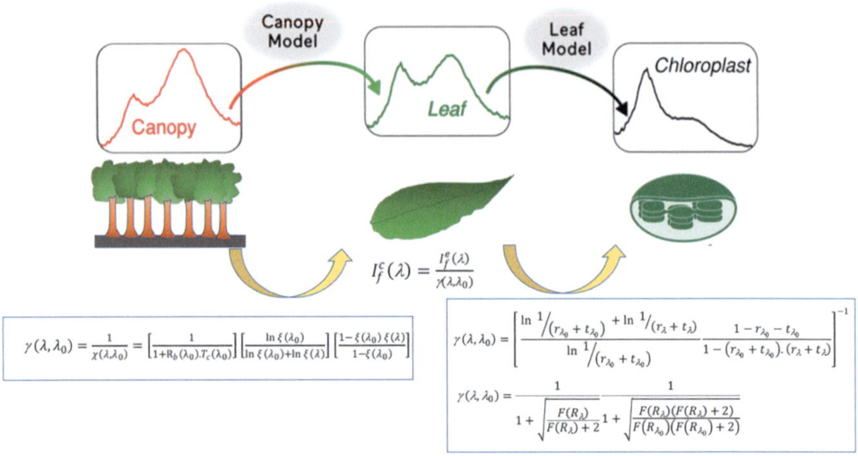

Fig. 6 Pictorial description of the application of two physical models presented in this chapter to retrieve the spectral fluorescence distribution at chloroplast level from canopy data. In the left part of the picture, the function $\gamma(\lambda, \lambda_0)$ represents the escape-from-canopy probability of a photon emitted at wavelength λ when the canopy is excited at λ_0. In the right part of the picture, the function $\gamma(\lambda, \lambda_0)$ represents the escape-from-leaf probability of a photon emitted at wavelength λ when the leaf is excited at λ_0. Two expressions are possible for this last calculation according to [4] or [3], respectively. Adapted from [31] with permission from Anales de la Asociación Química Argentina

In turn, $I_a(\lambda_0)$ can be written as the product of the incident photon flux $I_0(\lambda_0)$ and the fraction of light absorbed by the system f_a. The latter takes different expressions depending on the system. In general, $f_a = 1 - T_c(\lambda_0) - R(\lambda_0)$, which for an opaque system reduces to: $f_a = 1 - R(\lambda_0)$ [32].

Some examples of ϕ_{obs} calculation for defined systems and conditions are shown in Table 2.

Although the models presented in this chapter correspond to monochromatic excitation, Table 2 includes the expressions for the observed fluorescence quantum yields under polychromatic excitation.

General descriptions on the experimental determination of fluorescence quantum yields in scattering media, for different types of samples, can be found in [33].

5 About the Correction Function γ (λ, λ_0)

To conceptually and intuitively understand the fluorescence re-absorption effect, beyond the mathematical equations provided by the models, it is important to ponder its moderately complex dependence on emission and excitation wavelength.

When light absorption is high at the emission wavelength (generally associated with low reflectance values), fluorescence re-absorption is important, and consequently, the probability of photon escape given by $\gamma(\lambda, \lambda_0)$ is low (Fig. 7) and the effect of the correction is large.

Table 2 Summary of equations for observed quantum yield calculations

System	Excitation	Equation
Single leaf 	Monochromatic	$\phi_{\text{obs}} = \frac{\int_{600}^{800} I_{\text{F}}(\lambda)^{\text{exp}}.d\lambda}{[1 - T_{\text{L}}(\lambda_0) - R(\lambda_0)]I_0(\lambda_0)}$
Single leaf 	Polychromatic	$\phi_{\text{obs}} = \frac{\int_{600}^{800} I_{\text{F}}(\lambda)^{\text{exp}}.d\lambda}{\int_{400}^{800} [1 - T_{\text{L}}(\lambda_0) - R(\lambda_0)].I_0(\lambda_0).d\lambda_0}$
Thick layer of leaves 	Monochromatic	$\phi_{\text{obs}} = \frac{\int_{600}^{800} I_{\text{F}}(\lambda)^{\text{exp}}.d\lambda}{[1 - R(\lambda_0)]I_0(\lambda_0)}$
Thick layer of leaves 	Polychromatic	$\phi_{\text{obs}} = \frac{\int_{600}^{800} I_{\text{F}}(\lambda)^{\text{exp}}.d\lambda}{\int_{400}^{800} [1 - R(\lambda_0)].I_0(\lambda_0).d\lambda_0}$
Canopy 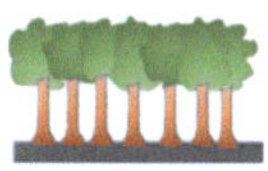	Monochromatic	$\phi_{\text{obs}} = \frac{\int_{600}^{800} I_{\text{F}}(\lambda)^{\text{exp}}.d\lambda}{[1 - T_{\text{c}}(\lambda_0) - R(\lambda_0)]I_0(\lambda_0)}$
Canopy 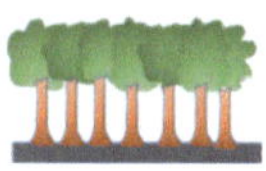	Polychromatic	$\phi_{\text{obs}} = \frac{\int_{600}^{800} I_{\text{F}}(\lambda)^{\text{exp}}.d\lambda}{\int_{400}^{800} [1 - T_{\text{c}}(\lambda_0) - R(\lambda_0)].I_0(\lambda_0).d\lambda_0}$

On the other hand, the absorption of light at the excitation wavelength works in the opposite direction (Fig. 8). The higher the absorption at the excitation wavelength, the lower the penetration of the beam into the sample, so that the fluorescence is generated at the surface layers of the system and is poorly reabsorbed. Conversely, when absorption at the excitation wavelength is weak, the light penetrates very deeply, the fluorescence is generated in the interior of the sample, and it travels a long distance inside the system before reaching an external detector and will consequently undergo high re-absorption on its way.

Then, to minimise light re-absorption, absorption should be high at the excitation wavelength and low at the emission wavelength. In such case, $\gamma(\lambda, \lambda_0)$ tends to unity.

In the case of extreme re-absorption (very low absorption at the excitation wavelength and very high absorption at the emission wavelength), $\gamma(\lambda, \lambda_0)$ tends to zero.

6 Final Remarks

To conclude, when studying the stationary fluorescence of chlorophyll-a, two key aspects are of primary interest: the spectral shape and the total intensity. Determining the "true" spectral shape (free from light re-absorption artefacts) is essential for

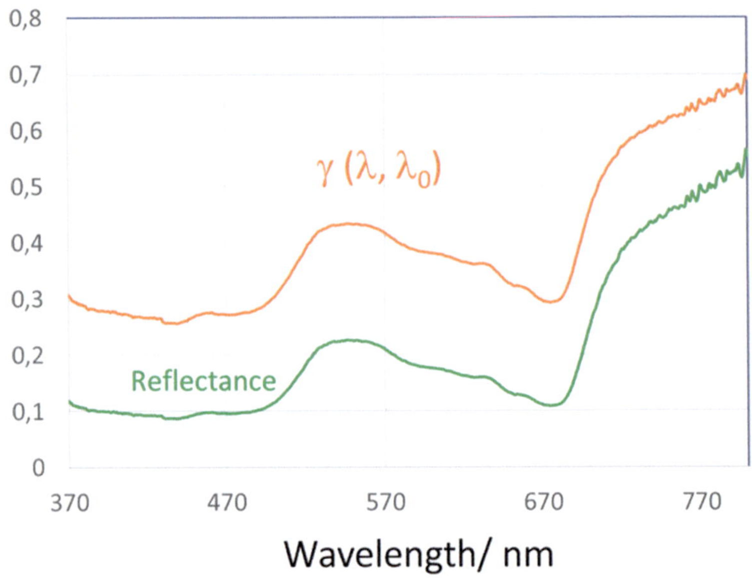

Fig. 7 Typical spectral distributions for the reflectance spectra of green leaves and the correction function $\gamma(\lambda, \lambda_0)$. In this graph, the excitation wavelength is kept constant ($\lambda_0 = 450$ nm) and $\gamma(\lambda, \lambda_0)$ is plotted against the emission wavelength

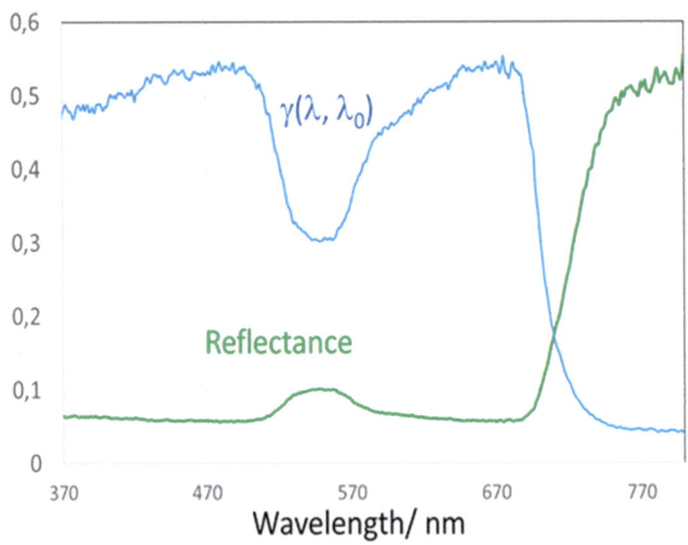

Fig. 8 Typical spectral distributions for the reflectance spectra of green leaves and the correction function $\gamma(\lambda, \lambda_0)$. In this graph, the emission wavelength ($\lambda = 680$ nm) is kept constant and $\gamma(\lambda, \lambda_0)$ is plotted against the excitation wavelength

obtaining fluorescence ratios that are directly related to plant physiology. This is achieved in the discussed models by dividing the experimental fluorescence spectrum by the function $\gamma(\lambda, \lambda_0)$. However, this quotient does not provide the corrected total fluorescence intensity as it does not take into account the infinite steps of re-absorption and re-emission of light. To account for these steps, one must calculate the corrected quantum yield and then multiply by the total absorbed photon flux Ia [31].

Moreover, to obtain the absolute "true" fluorescence spectrum, it is necessary to further multiply the corrected fluorescence intensity by the "true" spectral distribution $g(\lambda)$.

At room temperature, the fluorescence ratio ($F_{red}/F_{far\text{-}red}$) corrected for light re-absorption can change if the electron transport between PSII and PSI is blocked or damaged. Without stressors affecting the photosynthetic electron chain, the fluorescence ratio is expected to reflect photosystem stoichiometry. Various studies support this expectation [24, 34, 35]. In fact, it is well established that when chloroplasts are illuminated with longer wavelengths, which are preferentially absorbed by PSI, the amount of PSII and/or its antenna size increases to balance energy absorption between the two photosystems. This adaptation is manifested as a rise in the "true" or corrected fluorescence ratio ($F_{red}/F_{far\text{-}red}$), as shown in [24].

Finally, these corrections, which apply to stationary fluorescence spectra, should not be necessary for the analysis of variable chlorophyll fluorescence because in such methodologies, photosynthetic parameters are derived as ratios of fluorescence data measured at the same excitation and emission wavelengths. In fact, the correction factor cancels out in all the photosynthetic parameters that are calculated as relative quantities (ratios).

Important Summary Points

The model to be applied for light re-absorption processes is **dependent on the geometry** used for fluorescence measurements.

The approaches described in this chapter are valid when **fluorescence detection and excitation occur on the same side**.

The influence of fluorescence emission in reflectance and transmittance values was neglected in all cases.

These models are usually applied to chlorophyll-a fluorescence under **stationary conditions**.

All the approaches presented in this chapter involve **monochromatic excitation**.

Corrections for light re-absorption are not necessary when estimating photosynthetic parameters from variable chlorophyll fluorescence.

Appendix

Leaf Level: Corrections for Light Re-absorption and Scattering Processes

For a detailed derivation of equations used in the model described by Agati et al. 1993, you can refer to the original manuscript [4]. We will derive here the equations used in the work by Lagorio et al. 1998 [19].

We consider a system composed of a series of overlapping plant leaves, characterised by an absorption coefficient $k(\lambda)$ and a scattering coefficient $s(\lambda)$. The layer of leaves has a total depth d, and it is illuminated normally to the surface. To avoid edge effects, an infinite area is considered. The system is assumed to behave as an ideal homogeneous material for both the absorption and scattering of light, so that the irradiation is isotropic within the system. Additionally, the system is optically dense and no light is transmitted through it. Following the Kubelka-Munk theory, which describes the light behaviour in scattering media, the photon fluxes are described in a simplified manner by considering only two fluxes: one in the direction of the incident radiation and the other in the opposite direction (direction of reflected light) [7] (Fig. 9).

The attenuation of the fluxes is given by Eqs. (21) and (22).

$$\frac{\mathrm{d}i^{(0)}}{\mathrm{d}x} = -(k+s)\cdot i^{(0)} + s\cdot j^{(0)} \tag{21}$$

$$-\frac{\mathrm{d}j^{(0)}}{\mathrm{d}x} = -(k+s)\cdot j^{(0)} + s\cdot i^{(0)} \tag{22}$$

where $i^{(0)}$ and $j^{(0)}$ are the photon flows in the positive and negative directions, respectively. Solving (21) and (22) for monochromatic light and for a non-transmitting system, an expression for the Remission function $F(R_{\lambda 0})$ is obtained [7]:

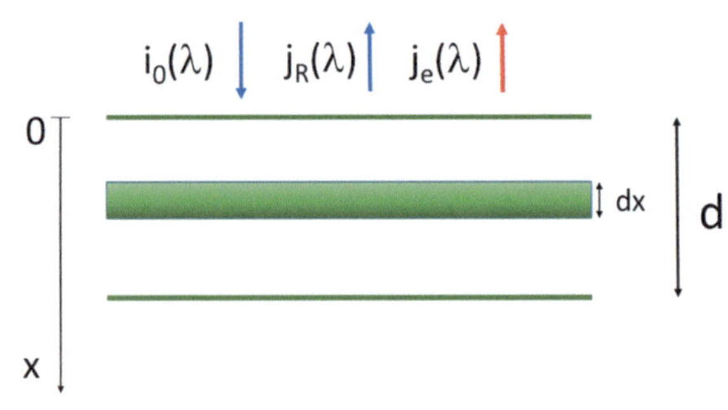

Fig. 9 Photon fluxes in the two-flux model

$$\frac{k(\lambda_0)}{s(\lambda_0)} = \frac{(1 - R_{\lambda 0})^2}{2R_{\lambda 0}} = F(R_{\lambda 0}) \tag{23}$$

In Eq. (23) $R_{\lambda 0}$ is the diffuse reflectance (quotient between the reflected flux at the front surface and the incident flux) at the excitation wavelength.

By introducing a fluorophore into the system, its emission (produced in each element volume) will be added to the existing light fluxes. We consider that this emission can also be decomposed into two photon fluxes: ($i^{(1)}$ and $j^{(1)}$) that suffer in turn an attenuation given by Eqs. (24) and (25):

$$\frac{\mathrm{d}i^{(1)}}{\mathrm{d}x} = -(k+s)\cdot i^{(1)} + s\cdot j^{(1)} + \frac{1}{2}g\cdot\varphi\frac{\mathrm{d}I_a^{(0)}}{\mathrm{d}x} \tag{24}$$

$$-\frac{\mathrm{d}j^{(1)}}{\mathrm{d}x} = -(k+s)\cdot j^{(1)} + s\cdot i^{(1)} + \frac{1}{2}g\cdot\varphi\frac{\mathrm{d}{I_a}^{(0)}}{\mathrm{d}x} \tag{25}$$

where $\frac{\mathrm{d}I_a^0}{\mathrm{d}x}$ is the differential photon flow absorbed by the emitting species, ϕ is the fluorescence quantum yield, and g is the emission spectrum which was normalised to area $=$ 1. Both ϕ and g represent quantities free from light re-absorption effects. Integration of Eqs. (24) and (25) with the boundary conditions: $i^{(1)}$ $(\lambda, x = 0) = 0$ and $j^{(1)}$ $(\lambda, x = \infty) = 0$ leads to the primary re-emitted radiation flow given by Eq. (26):

$$j^{(1)}(\lambda, x=0) = \varphi\cdot g(\lambda)\cdot I_0.(1 - R_{\lambda_0})\cdot\gamma(\lambda, \lambda_0) \tag{26}$$

where $\gamma(\lambda, \lambda_0)$ is described by Eq. (27)

$$\gamma(\lambda, \lambda_0) = \frac{1}{1+\sqrt{\frac{F(R_\lambda)}{F(R_\lambda)+2}}}\,\frac{1}{1+\sqrt{\frac{F(R_\lambda)\cdot(F(R_\lambda)+2)}{F(R_{\lambda_0})\cdot(F(R_{\lambda_0})+2)}}} \tag{27}$$

where λ and λ_0 represent the emission and excitation wavelengths, respectively.

So, the ratio

$$\frac{j^{(1)}(\lambda, x=0)}{\gamma(\lambda, \lambda_0)} = \varphi\cdot f(\lambda)\cdot I_0.(1 - R_{\lambda_0}) \tag{28}$$

shows the spectral distribution of fluorescence free from light re-absorption effects.

Canopy Level: Corrections for Light Re-absorption and Scattering Processes

The following derivation was originally described in [22] and [31]. In the canopy-soil system, the incident photon flux is partially reflected and partially absorbed (transmitted light is null). Part of the absorbed light is absorbed by the canopy and part by the ground. The following balances can therefore be considered (Eqs. (29) and (30)):

$$I_0 = I_\mathrm{R} + I_\mathrm{a} \tag{29}$$

and

$$I_\mathrm{a} = I_{ac} + I_\mathrm{ab}, \tag{30}$$

with I the photon flux, 0 means incident, a absorbed, c refers to canopy, and b to the background.

The photon flux absorbed by the background (I_ab) may be calculated by Eq. (31), where it was supposed that only a fraction of the land (C_f) is covered by the canopy. To derive this equation, it was considered that this photon flux results from the difference between the total number of incident photons and the number of reflected ones. Multi-scattering between the background (soil) and the canopy was taken into account (Fig. 10):

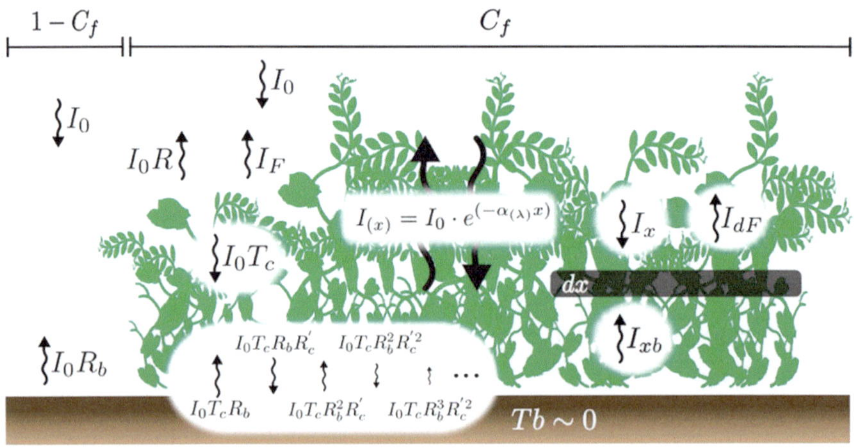

Fig. 10 Schematic representation of light beams interacting with a canopy. Multi-scattering between the soil and the canopy is pictorially described at the bottom left-hand part of the figure. (Reprinted from Romero et al. [21], Copyright (2020), with permission from Elsevier)

$$I_{ab} = (1 - C_f)(I_0 - I_0 R_b) + C_f\left(I_0 T_c + I_0 T_c R_b R_c^{'} + I_0 T_c R_b^2 R_c^{'2} + \ldots\right) \\ - C_f\left(I_0 T_c R_b + I_0 T_c R_b^2 R_c^{'2} + I_0 T_c R_b^3 R_c^{'2} + \ldots\right),$$

$$I_{ab} = I_0(1 - C_f)(1 - R_b) + C_f[I_0 T_c](1 + r + r^2 + \ldots) - C_f[I_0 T_c R_b](1 + r + r^2 + \ldots)$$

And

$$\mathrm{I}_{ab} = I_0(1 - C_f)(1 - R_b) + C_f[I_0 T_c]\left(\frac{1}{1 - r}\right) - C_f[I_0 T_c R_b]\left(\frac{1}{1 - r}\right), \tag{31}$$

where I_{Tc} denotes the amount of light transmitted by the canopy, I_0 represents the incident photon flux, R_b is the background reflectance, and R'_c refers to the upward reflectance of the canopy.

For the area devoid of plants, I_0 represents the incident light, $I_0.R_b$, the reflected light, and the difference between them gives the number of absorbed photons. Since 1-C_f represents the fraction of land lacking vegetation, the initial term in Eq. (31) represents the contribution of the bare land to the total absorbed photons. The fraction of soil covered by plants (C_f) receives $I_0.T_c$ photons. A portion of these $I_0 \cdot T_c \cdot R_b$ is reflected by the background and subsequently, $I_0 \cdot T_c \cdot R_b \cdot R_c'$ is reflected from the canopy to the soil. Consequently, the soil reflects $I_0 \cdot T_c \cdot R^2{}_b \cdot R_c'$ and this process continues iteratively. Hence, the contribution of land with vegetation to the total absorbed photons is contained in the second and third terms in Eq. (31).

In this equation, the product $R_b R_c'$ was substituted by r ($r < 1$), and the sum of the infinite geometric sequence ($1 + r + r^2 + \ldots$) was replaced with $1/(1 - r)$.

Equation (32) was then obtained by substituting Eqs. (30) and (31) in (29) and dividing by I_0,

$$1 = R + a_c + (1 - C_f)(1 - R_b) + C_f \cdot T_c \cdot (1 - R_b) \cdot \left(\frac{1}{1 - r}\right) \tag{32}$$

where R represents the total reflectance of the system, a_c is the proportion of light absorbed by the canopy, and T_c denotes the transmittance through the canopy. Equation (33) specifies the fraction of photons that the canopy does not absorb (Fig. 11).

$$1 - a_c = R + (1 - C_f)(1 - R_b) + C_f \cdot T_c \cdot (1 - R_b) \cdot \left(\frac{1}{1 - r}\right) \tag{33}$$

In the following, we analyse the light absorption and emission processes considering the system divided into differential layers. The excitation light and the light reflected by the soil that reach a layer are assumed to be monoexponentially attenuated (Eqs. (34) and (35)) (Fig. 12).

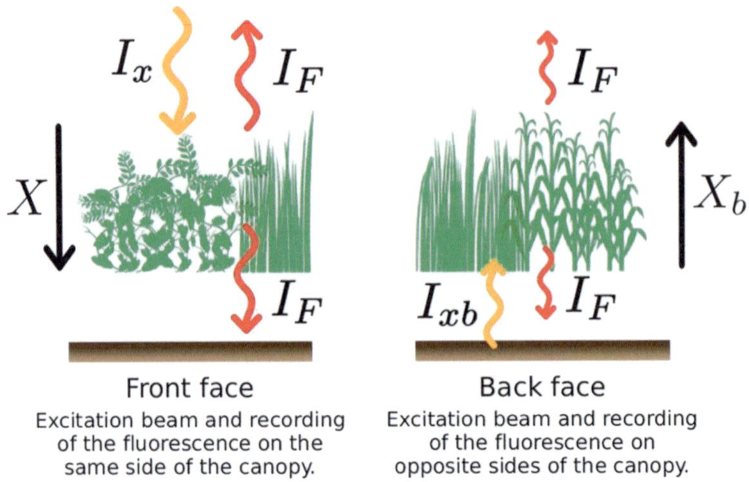

Fig. 11 Schematic description of the differences in light absorption for the front-face and back-face events. Front face: input of the excitation beam and recording of fluorescence on the same side of the canopy, back face: input of the excitation and recording of fluorescence on opposite sides of the canopy. (Reprinted from Romero et al. [21], Copyright (2020), with permission from Elsevier)

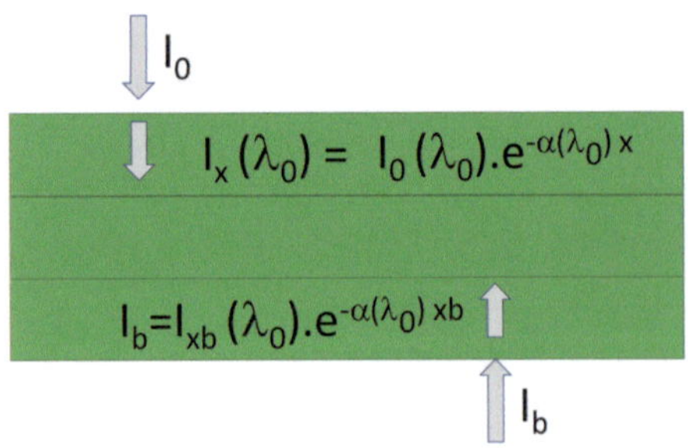

Fig. 12 Attenuation of the incident beam and of the beam reflected by the soil when reaching a differential layer

$$I_x(\lambda_0) = I_0(\lambda_0).e^{-\alpha(\lambda_0).x(\lambda_0)} \tag{34}$$

and

$$I_{xb}(\lambda_0) = I_b(\lambda_0).e^{-\alpha(\lambda_0).x_b(\lambda_0)} \tag{35}$$

where λ_0 represents the excitation wavelength, $\alpha(\lambda)$ the absorption coefficient of the canopy, and $x(\lambda_0)$ and $x_b(\lambda_0)$ are the effective path lengths at λ_0 that are affected by light scattering.

Traversing the layer, the beams suffer additional attenuation (Fig. 13).

The difference between the number of photons that enter and those that exit gives the photon flux absorbed by the layer ($A_{dx}(\lambda_0)$):

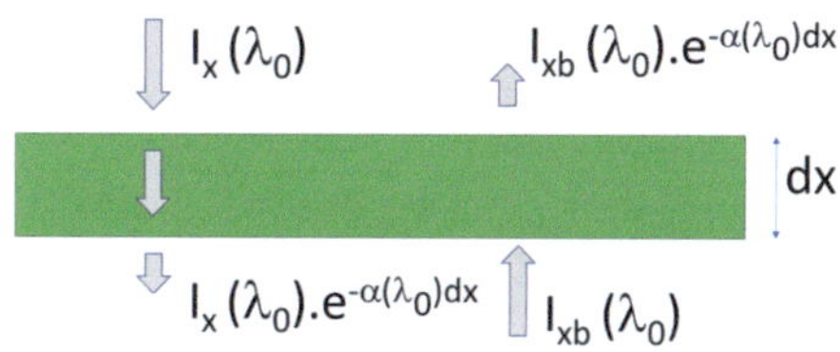

Fig. 13 Traversing the layer, the beams suffer additional attenuation

$$A_{dx}(\lambda_0)=I_x(\lambda_0)+I_{xb}(\lambda_0)-\left[I_x(\lambda_0).e^{-\alpha(\lambda_0).dx}+I_{xb}(\lambda_0).e^{-\alpha(\lambda_0).dx}\right],$$
$$A_{dx}(\lambda_0)=I_x(\lambda_0)\left[1-e^{-\alpha(\lambda_0).dx}\right]+I_{xb}(\lambda_0)\left[1-e^{-\alpha(\lambda_0).dx}\right],$$
$$A_{dx}(\lambda_0)=\left[I_x(\lambda_0)+I_{xb}(\lambda_0)\right]\left[1-e^{-\alpha(\lambda_0).dx}\right] \quad (36)$$

The absorbed photon flux by the differential layer generates a fluorescence differential ($dI_F(\lambda)$) which is a function of wavelength (Eq. (37))

$$dI_F(\lambda)=\phi_F.g(\lambda).A_{dx}(\lambda_0),$$
$$dI_F(\lambda)=\phi_F.g(\lambda).\left[I_x(\lambda_0)+I_{xb}(\lambda_0)\right]\left[1-e^{-\alpha(\lambda_0).dx}\right] \quad (37)$$

In Eq. (37), ϕ_F is the number of photons emitted per absorbed photon and $g(\lambda)$ is the spectral distribution of fluorescence normalised to unity area.

Introducing (34) and (35) in (37), with the assumptions that ϕ_F is the same for all the layers and that $\lim_{w\to 0} 1-e^{w}=-w$, Eq. (38) is derived.

$$dI_F(\lambda)=\phi_F.g(\lambda).\alpha(\lambda_0).\left\{I_0(\lambda_0).\left[e^{-\alpha(\lambda_0).x(\lambda_0)}\right]+I_b(\lambda_0).\left[e^{-\alpha(\lambda_0).x_b(\lambda_0)}\right]\right\}dx \quad (38)$$

Now, to consider that this fluorescence differential is attenuated by light-re-absorption process within the canopy, Eq. (38) is multiplied by $e^{-\alpha(\lambda).\ x(\lambda)}$ and the experimental value for this differential is obtained (Eq. (39)).

$$dI_F(\lambda)^{exp}$$
$$=\phi_F.g(\lambda).\alpha(\lambda_0).\left\{I_0(\lambda_0).\left[e^{-\alpha(\lambda_0).x(\lambda_0)}\right]+I_b(\lambda_0).\left[e^{-\alpha(\lambda_0).x_b(\lambda_0)}\right]\right\}.e^{-\alpha(\lambda).x(\lambda)}dx \quad (39)$$

It should be noted that the attenuation factor ($e^{-\alpha(\lambda).\ x(\lambda)}$) introduced at this step depends on the emission wavelength unlike the previous factors which depended on the excitation wavelengths.

Now, let us focus on the variable $x(\lambda)$ and its dependence on the wavelength of light. To introduce the effect of scattering in the canopy, we can rewrite $x(\lambda)$ as

$$x(\lambda)=Z{\cdot}f(\lambda) \quad (40)$$

where Z is the maximum average effective optical path length at the excitation wavelength and the function $f(\lambda)$ contains the wavelength dependence of the scattering coefficient.

To obtain a theoretical expression for the fluorescence intensity generated by the canopy, integration over the whole system is performed. Equation (40) is replaced into Eq. (39), $x_b(\lambda_o)$ is expressed as $x_b(\lambda_o) = Z - x(\lambda_0)$, and integration of both Eqs. (38) and (39) between $x = 0$ and $x = Z$ is carried out. This procedure leads to Eq. (41) for the fluorescence free of re-absorption and (42) for the experimental fluorescence:

$$I_F(\lambda) = \int_0^Z \phi_F.g(\lambda).\alpha(\lambda_0).[I_0(\lambda_0).e^{-\alpha(\lambda_0).x(\lambda_0)} + I_b(\lambda_0).e^{-\alpha(\lambda_0).(Z-x(\lambda_0))}]dx, \quad (41)$$

$$I_F(\lambda)^{exp} = \int_0^Z \phi_F.g(\lambda).\alpha(\lambda_0).[I_0(\lambda_0).e^{-\alpha(\lambda_0).x(\lambda_0)} + I_b(\lambda_0).e^{-\alpha(\lambda_0).(Z-x(\lambda_0))}].e^{-\alpha(\lambda).x(\lambda_0).f(\lambda)}dx \quad (42)$$

Solving Eqs. (41) and (42) results in Eqs. (43) and (44), respectively,

$$I_F(\lambda)^{corr} = \phi_F.g(\lambda).\alpha(\lambda_0).[I_0(\lambda_0) + I_b(\lambda_0)]\ [1 - e^{-\alpha(\lambda_0).Z}], \quad (43)$$

$$I_F(\lambda)^{exp} = \phi_F.g(\lambda).\left\{ \begin{array}{l} I_0(\lambda_0)[\frac{\alpha(\lambda_0)}{\alpha(\lambda_0)+\alpha(\lambda).f(\lambda)}][1 - e^{-\alpha(\lambda_0).Z-\alpha(\lambda).f(\lambda).Z}] \\ +I_b(\lambda_0)[\frac{\alpha(\lambda_0)}{\alpha(\lambda).f(\lambda)-\alpha(\lambda_0)}][e^{-\alpha(\lambda_0).Z} - e^{-\alpha(\lambda).f(\lambda).Z}] \end{array} \right\} \quad (44)$$

where $I_F(\lambda)$ in the absence of light re-absorption processes was renamed as $I_F(\lambda)^{corr}$.

From Eqs. (43) and (44), it follows that the fluorescence spectrum free from light re-absorption artefacts, $I_F(\lambda)^{corr}$, can be retrieved from the experimental spectrum by multiplying the latter by the factor $\chi(\lambda,\lambda_0)$ given by Eq. (45).

$$\chi(\lambda,\lambda_0) = \frac{I_F(\lambda)^{corr}}{I_F(\lambda)^{exp}} = \frac{[I_0(\lambda_0)+I_b(\lambda_0)][1-e^{-\alpha(\lambda_0).Z}]}{I_0(\lambda_0)[\frac{\alpha(\lambda_0)}{\alpha(\lambda_0)+\alpha(\lambda).f(\lambda)}][1-e^{-\alpha(\lambda_0).Z-\alpha(\lambda).f(\lambda).Z}]+I_b(\lambda_0)[\frac{\alpha(\lambda_0)}{\alpha(\lambda).f(\lambda)-\alpha(\lambda_0)}][e^{-\alpha(\lambda_0).Z}-e^{-\alpha(\lambda).f(\lambda).Z}]} \quad (45)$$

For practical purposes, it is convenient to rewrite Eq. (45) in terms of the fraction of light non-absorbed by the canopy. The effective average path length as a function of wavelength $Z(\lambda)$ can be calculated as the product $Z.\ f(\lambda)$, according to Eq. (40). The effective absorption spectrum of the canopy is then obtained by multiplying this path length by the absorption coefficient $\alpha(\lambda).\ Z(\lambda)$. Consequently, Eq. (46) can be derived for the fraction of light non-absorbed by the canopy ($\xi(\lambda)$):

$$\xi(\lambda) = \mathrm{e}^{-\alpha(\lambda).Z(\lambda)} \tag{46}$$

Once this expression is known, the correction factor can be rewritten in terms of $\xi(\lambda)$ (Eq. (47)):

$$\chi(\lambda,\lambda_0) = \frac{[I_0(\lambda_0)+I_\mathrm{b}(\lambda_0)]\,[1-\xi(\lambda_0)]}{I_0(\lambda_0)[\frac{\ln\xi(\lambda_0)}{\ln\xi(\lambda_0)+\ln\xi(\lambda)}][1-\xi(\lambda_0)\,\xi(\lambda)]+I_\mathrm{b}(\lambda_0)[\frac{\ln\xi(\lambda_0)}{\ln\xi(\lambda)-\ln\xi(\lambda_0)}][\xi(\lambda_0)-\xi(\lambda)]} \tag{47}$$

The inverse of $\chi(\lambda,\lambda_0)$ (48) corresponds to the function $\gamma(\lambda,\lambda_0)$ for the canopy. This function, representing the escape probability of a photon emitted at λ when the sample is excited at λ_0, was introduced generally in Eq. (4) and takes diverse mathematical forms for each system under study.

$$\begin{aligned}\gamma(\lambda,\lambda_0) &= \frac{1}{\chi(\lambda,\lambda_0)}\\ &= \left[\frac{I_0(\lambda_0)}{[I_0(\lambda_0)+I_\mathrm{b}(\lambda_0)]}\right]\left[\frac{\ln\xi(\lambda_0)}{\ln\xi(\lambda_0)+\ln\xi(\lambda)}\right]\left[\frac{1-\xi(\lambda_0)\,\xi(\lambda)}{1-\xi(\lambda_0)}\right]\\ &+\left[\frac{I_\mathrm{b}(\lambda_0)}{[I_0(\lambda_0)+I_\mathrm{b}(\lambda_0)]}\right]\left[\frac{\ln\xi(\lambda_0)}{\ln\xi(\lambda)-\ln\xi(\lambda_0)}\right]\left[\frac{\xi(\lambda_0)-\xi(\lambda)}{1-\xi(\lambda_0)}\right]\end{aligned} \tag{48}$$

The advantage of having an expression for the correction factor in terms of $\xi(\lambda)$ is that this factor can be easily calculated from experimental data as $R+(1-C_\mathrm{f})\times(1-R_\mathrm{b})+C_\mathrm{f}\cdot T_\mathrm{c}\cdot(1-R_\mathrm{b})\cdot\left(\frac{1}{1-r}\right)$ (Eq. (33)).

Additionally, the factor $I_\mathrm{b}(\lambda_0)$ is correlated with the experimental $R_\mathrm{b}(\lambda_0)$ and $T_\mathrm{c}(\lambda_0)$ (Eq. (49)).

$$I_\mathrm{b}(\lambda_0) = R_\mathrm{b}(\lambda_0).T_\mathrm{c}(\lambda_0).I_0(\lambda_0) \tag{49}$$

Replacing (49) in (48) provides the final expression for the correction factor (50) for the canopy.

$$\begin{aligned}\gamma(\lambda,\lambda_0) &= \frac{1}{\chi(\lambda,\lambda_0)}\\ &= \left[\frac{1}{1+R_\mathrm{b}(\lambda_0).T_\mathrm{c}(\lambda_0)}\right]\left[\frac{\ln\xi(\lambda_0)}{\ln\xi(\lambda_0)+\ln\xi(\lambda)}\right]\left[\frac{1-\xi(\lambda_0)\,\xi(\lambda)}{1-\xi(\lambda_0)}\right]\\ &+\left[\frac{R_\mathrm{b}(\lambda_0).T_\mathrm{c}(\lambda_0)}{1+R_\mathrm{b}(\lambda_0).T_\mathrm{c}(\lambda_0)}\right]\left[\frac{\ln\xi(\lambda_0)}{\ln\xi(\lambda)-\ln\xi(\lambda_0)}\right]\left[\frac{\xi(\lambda_0)-\xi(\lambda)}{1-\xi(\lambda_0)}\right]\end{aligned} \tag{50}$$

References

1. Cordon GB, Lagorio MG (2006) Re-absorption of chlorophyll fluorescence in leaves revisited. A comparison of correction models. Photochem Photobiol Sci 5:735–740. https://doi.org/10.1039/B517610G
2. Cendrero-Mateo MP, Moran MS, Papuga SA, Thorp KR, Alonso L, Moreno J, Ponce-Campos-G, Rascher U, Wang G (2016) Plant chlorophyll fluorescence: active and passive measurements at canopy and leaf scales with different nitrogen treatments. J Exp Bot 67:275–286. https://doi.org/10.1093/jxb/erv456
3. Ramos ME, Lagorio MG (2004) True fluorescence spectra of leaves. Photochem Photobiol Sci 3:1063–1066. https://doi.org/10.1039/b406525e
4. Agati G, Fusi F, Mazzinghi P, Lipucci di Paola M (1993) A simple approach to the evaluation of the reabsorption of chlorophyll fluorescence spectra in intact leaves. J Photochem Photobiol B Biol 17:163–171. https://doi.org/10.1016/1011-1344(93)80009-X
5. Gitelson AA, Buschmann C, Lichtenthaler HK (1998) Leaf chlorophyll fluorescence corrected for re-absorption by means of absorption and reflectance measurements. J Plant Physiol 152: 283–296. https://doi.org/10.1016/S0176-1617(98)80143-0
6. Lichtenthaler HK, Hak R, Rinderle U (1990) The chlorophyll fluorescence ratio F690/F730 in leaves of different chlorophyll content. Photosynth Res 25:295–298. https://doi.org/10.1007/BF00033170
7. Wendlandt WWM, Hecht HG (1966) Reflectance spectroscopy. Intersciences Publishers, New York
8. Iriel A, Dundas G, Fernández Cirelli A, Lagorio MG (2015) Effect of arsenic on reflectance spectra and chlorophyll fluorescence of aquatic plants. Chemosphere 119:697–703. https://doi.org/10.1016/j.chemosphere.2014.07.066
9. Iriel A, Mendes Novo J, Cordon GB, Lagorio MG (2014) Atrazine and methyl viologen effects on chlorophyll-a fluorescence revisited. Implications in photosystems emission and ecotoxicity assessment. Photochem Photobiol 90:107–112. https://doi.org/10.1111/php.12142
10. Peterson RB, Oja V, Laisk A (2001) Chlorophyll fluorescence at 680 and 730 nm and leaf photosynthesis. Photosynth Res 70:185–196. https://doi.org/10.1023/A:1017952500015
11. Agati G, Cerovic ZG, Moya I (2000) The effect of decreasing temperature up to chilling values on the *in vivo* F685/F735 chlorophyll fluorescence ratio in *Phaseolus vulgaris* and *Pisum sativum*: the role of the photosystem I contribution to the 735 nm fluorescence band. Photochem Photobiol 72:75–84. https://doi.org/10.1562/0031-8655(2000)072<0075:teodtu>2.0.co;2
12. Agati G (1998) Response of the *in vivo* chlorophyll fluorescence spectrum to environmental factors and laser excitation wavelengths. Pure Appl Opt 7:797–807. https://doi.org/10.1088/0963-9659/7/4/016
13. Buschman C, Lichtenthaler HK (1998) Principles and characteristics of multi-colour fluorescence imaging of plants. J Plant Physiol 152:297–314
14. Porcar-Castell A, Tyystjärvi E, Atherton J, van der Tol C, Flexas J, Pfündel EE, Moreno J, Frankenberg C, Berry JA (2014) Linking chlorophyll a fluorescence to photosynthesis for remote sensing applications: mechanisms and challenges. J Exp Bot 65:4065–4095. https://doi.org/10.1093/jxb/eru191
15. Birks JB (1976) Fluorescence quantum yield measurements. J Res Natl Bur Stand Sect A 80A:389–399. https://doi.org/10.6028/jres.080A.038
16. Birks JB (1974) Fluorescence parameters and their interpretation. J Lumin 9:311–314. https://doi.org/10.1016/0022-2313(74)90044-1
17. Birks JB (1954) Energy transfer in organic phosphors. Phys Rev 94:1567–1573. https://doi.org/10.1103/PhysRev.94.1567
18. Rodriguez HB, Mirenda M, Lagorio MG, Román S (2019) Acc Chem Res 52:110–118. https://doi.org/10.1021/acs.accounts.8b00271

19. Lagorio MG, Dicelio LE, Litter MI, San Román E (1998) Modeling of fluorescence quantum yields of supported dyes. Aluminium carboxyphthalocyanine on cellulose. J Chem Soc Faraday Trans 94:419–425. https://doi.org/10.1039/A706113G
20. Braslavsky SE (2007) Glossary of terms used in photochemistry, 3rd edition (IUPAC recommendations 2006). Pure Appl Chem 79:293. https://doi.org/10.1351/pac200779030293
21. Romero JM, Cordon GB, Lagorio MG (2020) Re-absorption and scattering of chlorophyll fluorescence in canopies: a revised approach. Remote Sens Environ 246:111860. https://doi.org/10.1016/j.rse.2020.111860
22. Romero JM, Cordon GB, Lagorio MG (2018) Modeling re-absorption of fluorescence from the leaf to the canopy level. Remote Sens Environ 204:138–146. https://doi.org/10.1016/j.rse.2017.10.035
23. Cordon GB, Lagorio MG (2007) Absorption and scattering coefficients. A biophysical-chemistry experiment using reflectance spectroscopy. J Chem Educ 84:1167–1170. https://doi.org/10.1021/ed084p1167
24. Cordon GB, Lagorio MG (2007) Optical properties of the adaxial and abaxial faces of leaves. Chlorophyll fluorescence, absorption and scattering coefficients. Photochem Photobiol Sci 6: 873–882. https://doi.org/10.1039/B617685B
25. Olf HG (1988) Stokes's pile of plates revisited. J Opt Soc Am A 5:1620–1625. https://doi.org/10.1364/JOSAA.5.001620
26. Gausman HW, Allen WA (1973) Optical parameters of leaves of 30 plant species. Plant Physiol 52:57–62. https://doi.org/10.1104/pp.52.1.57
27. Allen WA, Richardson AJ (1968) Interaction of light with a plant canopy. Opt Soc Am 58: 1023–1028. https://doi.org/10.1364/JOSA.58.001023
28. Stokes GG (1860) On the intensity of the light reflected from or transmitted through a pile of plates. Proc R Soc Lond 11:545–556. http://www.jstor.org/stable/111970
29. Goedheer JC (1964) Fluorescence bands and chlorophyll a forms. Biochimica et Biophysica Acta (BBA) 88:304–317. https://doi.org/10.1016/0926-6577(64)90186-X
30. Louis J, Cerovic ZG, Moya I (2006) Quantitative study of fluorescence excitation and emission spectra of leaves. J Photochem Photobiol B 85:65–71. https://doi.org/10.1016/j.jphotobiol.2006.03.009
31. Romero JM, Torres R, Ospina Calvo B, Diz VE, Iriel A, Cordon GB, Lagorio MG (2020) Photochemistry and photophysics of biological systems, chlorophyll fluorescence and photosynthesis. An Asoc Qca Arg 107:1–32. https://www.aqa.org.ar/images/anales/pdf107-2/107-2.pdf
32. Lagorio MG, Cordon GB, Iriel A, Romero JM, Faivovich J, Taboada C (2020) Biophotonics. Fluorescence and reflectance in living organisms. Sci Rev Few 2:18–41. https://doi.org/10.52712/sciencereviews.v2i1.36
33. Lagorio MG (2020) Tutorial: determination of fluorescence quantum yields in scattering media. Methods Appl Fluoresc 8:043001. https://doi.org/10.1088/2050-6120/aba69c
34. Ospina Calvo B, Lagorio MG (2019) Quantitative effects of pigmentation on the re-absorption of chlorophyll a fluorescence and energy partitioning in leaves. Photochem Photobiol 95:1360–1368. https://doi.org/10.1111/php.13149
35. Ospina Calvo B, Parapugna TL, Lagorio MG (2017) Variability in chlorophyll fluorescence spectra of eggplant fruit grown under different light environments: a case study. Photochem Photobiol Sci 16:711–710. https://doi.org/10.1039/C6PP00475J

Fluorescence Applications on Plants

Giovanni Agati

Contents

Abstract Recent significant examples of fluorescence spectroscopy and imaging applications in plant science are reported. Particular emphasis has been placed on the non-destructive fluorescence detection of plant polyphenolic compounds and of plant stress indicators. These techniques could be fruitfully applied in agriculture to control plant diseases and stress conditions avoiding crop yield losses and to assess the quality of vegetable and fruit products.

The aim of this chapter is to stimulate the younger researchers in photobiology to better understand the advantages and limits of the different *in planta* fluorescence techniques available, for their correct applications.

The original version of the chapter has been revised. A correction to this chapter can be found at https://doi.org/10.1007/17361_2025_13

G. Agati (✉)
Institute of Applied Physics "Nello Carrara" (IFAC), National Research Council, Florence, Italy

Consortium INSTM-Italian Interuniversity Consortium for Science and Technology of Materials, Florence, Italy
e-mail: g.agati@ifac.cnr.it

M. G. Lagorio (ed.), Comprehensive Series in Photochemical and Photobiological Sciences, Fluorescence of Living Plants 1, https://doi.org/10.1007/17361_2024_3
Published online: 25 March 2025

Keywords Chlorophyll fluorescence excitation screening, CLSM fluorescence imaging, Detection of phenolic compounds, Disease detection, In vivo fluorescence probes, Micronutrient detection, Stress condition indices, Tissue compound localization, X-ray fluorescence spectroscopy

1 Introduction

Most of the applications of fluorescence spectroscopy in plant science concern the detection of chlorophyll (Chl) fluorescence and its strict link to the plant photosynthetic activity.

Uses of chlorophyll fluorescence (ChlF), as both induction kinetics and imaging analyses, to detect abiotic and biotic induced plant stress status have been comprehensively and recently reviewed [1–3]. Therefore, this issue does not need to be treated here, and the readers can refer to those mentioned reviews and the within reported literature to go into details on it.

Besides this, considering the in vivo chlorophyll fluorescence spectral shape, the ratio between the red and far-red (or vice versa) ChlF bands has been largely used to predict non-destructively the leaf chlorophyll content [4]. This is based on the partial reabsorption of the red emission band by Chl itself that is proportional to the leaf Chl concentration. The fluorescence approach has the advantage with respect to reflectance spectroscopy to be plant specific and then less sensitive to interference from other targets, especially in remote sensing.

Determination of the leaf chlorophyll content is also possible by retrieving the solar-induced ChlF ratio, F686.4/F760.5, in the two O_2A and O_2B oxygen absorption bands of terrestrial atmosphere (Fraunhofer line depth method) [5].

In vivo plant fluorescence detection can be performed over a wide range of scale from the microscopic sub-cellular level to the whole crop/forest level by using different techniques and devices. See for example the application of fluorescence lifetime imaging microscopy (FLIM) to identify the wood cell wall lignin composition in aspen trees and the dynamics of lignin biosynthesis in *Arabidopsis thaliana* roots [6]. Recently, a novel compact and low-cost multispectral fluorescence LiDAR, with a 405 nm diode laser excitation and an RGB camera as detector, was developed and proposed for high-throughput plant phenotyping [7]. The possibility of retrieving solar-induced fluorescence (SIF) from different remote sensing platforms (ground, UAV, airborne, and spaceborne) and its applications has been comprehensively reviewed by Bandopadhyay et al. (2020) [8].

Alternative use of ChlF consists of the indirect non-destructive estimation of the content of phytochemicals in the leaf epidermis and fruit skin. The emission signals at different excitation wavelengths by Chl in the plant organ tissues underneath the superficial layers are employed to calculate the in vivo absorbance of epidermal compounds [9].

Several other studies reported on the in vivo detection of other natural fluorescence molecules counting: (i) polyphenol compounds bound to leaf and fruit superficial structures; (ii) phytoalexins and markers of molecules involved in the mechanisms of response to pathogen infection; (iii) fluorescence probes of stress indicators; and (iv) plant micronutrients.

In this contribution, I am going to report some significant recent examples of fluorescence use in plant science, without having the aim to be exhaustive in treating a so vast and intriguing matter.

2 Chlorophyll Fluorescence Ratio

The ChlF ratio (ChlFR) is often defined as the ratio between the red and the far-red Chl emission bands and resulted to be inversely related to the leaf Chl content. More recently, the opposite far-red to red ChlFR was introduced and named the Simple Fluorescence Ratio (SFR) in order to have an index positively related to the plant Chl content. Changes in ChlFR, or SFR, under stress can be associated to either a change in Chl concentration or a change in the photosynthesis status of the plant, with the first contribution usually occurring on a longer term than the second.

ChlFR was found to be a useful parameter to monitor biomass during developing of basil plants under growth chamber and greenhouse conditions [10]. A linear relationship on a logarithm scale between ChlFR measured at the canopy level and the leaf dry weight was observed. The simple proximal spectrofluorometer system used along with the optimization of acquisition protocols and signal calibration indicate the possibility of a commercial application of the method.

Laser-induced fluorescence (LIF) was used to evaluate possible ChlF spectral signature of water stress in the model plant *A. thaliana* [11]. Measurements were acquired on potted plants during slowly applied water deficit for 12 days, as well as on detached leaves during drying at room temperature. In both cases, significant changes in the ChlFR as function of the leaf water content were detected. Interestingly, shifts in the wavelength emission maxima with water stress were also observed and can then be taken into account as an additional stress indicator. LIF techniques for the detection of plant drought stress can represent an additional tool to the conventional pulse amplitude modulation (PAM) fluorescence methods with the advantage of not requiring dark adaptation of the leaves and allowing the remote sensing application directly in the field.

3 Multispectral Fluorescence and Stress Detection

When plant green organs are excited by UV or short-blue radiation, besides ChlF, a blue-green emission can be detected. This originates mainly from polyphenolic compounds present in the cuticles or in the epidermal cell walls and vacuoles. The composition and concentration of the superficial polyphenolic compounds are sensitive to plant stress conditions; therefore, the detection of both blue-green and red-far-red emission bands can provide an additional signature of the plant's physiological status.

Usually, four different emission bands (under 320–405 nm excitation) located around the blue (F440), green (F520), red (F690), and far-red (F740) spectral regions are retrieved, and the relative F440/F520, F440/F690, F440/F740, and F690/F740 fluorescence ratios are considered for early stress detection applications.

Multispectral fluorescence imaging adds the information about the heterogeneity of the fluorescence signals on the sample surface and their pattern change with stresses. Early use of the technique showed the possibility to apply it for the sensing of several plant stresses as nitrogen deficiency, water deficiency, sun exposure, and mite attacks [12].

Multispectral fluorescence imaging was proposed for the early detection of infection by bacteria and fungi in *Cucurbita pepo* leaves [13], or to follow the occurrence of drought stress in *A. thaliana* [14].

Combination of the four blue, green, red, and far-red fluorescence images under 365 nm excitation along with principal component analysis and support vector machine classification could detect salt-stressed plants of *A. thaliana* with an accuracy of 92% after 5 days of salt treatment [15].

4 Detection of Plant Phenolic Compounds

Phenolic compounds are plant metabolites playing fundamental roles in counteracting abiotic and biotic stresses, as protectors from excess visible and UV radiation, reactive oxygen species (ROS) scavengers, and antimicrobials.

Their nutraceutical (antioxidants) properties add quality value to fruits and vegetables.

Therefore, in vivo detection of phenolic compounds can represent a useful tool to better understand their mechanisms of action in ecophysiology studies or to control the quality of crop products in precision agriculture applications.

As it will be shown below, some phenolic compounds possess significant in vivo fluorescence quantum yield so that they can be directly detected by fluorescence spectroscopy, depending on their localization in the organ tissues (superficial or inside) and the penetration into the plant cells and structures of the excitation wavebands.

4.1 Chlorophyll Fluorescence Excitation Screening Method

Non-fluorescent compounds located on the leaf epidermis or fruit skin can still be detected indirectly by using the ChlF excitation screening method. It simply takes into account the spectral attenuation of the ChlF excitation by compounds located in the plant external tissues, so that ChlF at specific excitation wavebands will be inversely related to the concentration of epidermal compounds.

The description of the technique and references of its applications in the agro-food and plant science sectors can be found in the literature [9, 16, 17].

The comparison of the in situ ChlF excitation spectra of adaxial and abaxial sides of bifacial leaves, or between older and younger segments of monocotyledonous leaves, allowed for the assignment of the type of UV-screening compounds present in the leaf epidermis [18, 19].

In-field applications of the method became possible by the development of dedicated portable optical sensors. In particular, green and red excitations of the far-red ChlF provided an index of leaf and fruit anthocyanins (ANTH), while UVA and red excitations of the far-red ChlF determined an index of leaf and fruit flavonoids (FLAV). The ANTH index has been extensively used to estimate the phenolic maturity in wine grape in the vineyards and mapping its spatial heterogeneity to control harvesting for the production of high-quality wines [20, 21].

The FLAV index, being inversely related to the leaf nitrogen content, was employed in several studies, along with a Chl index, to evaluate the plant nutrient status aimed to conduct crop precision fertilization [22, 23].

Interestingly, the FLAV index represented a valid tool to study the daily change in the flavonoid content of leaf epidermis as a rapid response of the plants to the daily variation of solar UV radiation intensity [17, 24].

4.2 Compound Localization

Knowing the tissue distribution of phenolic compounds is helpful to understand their main role in the plant response to stress conditions.

Confocal laser scanning microscopy (CLSM) can be a useful tool to non-invasively localize some classes of phenolic compounds inside plant organ tissues. It allows for the observation of both intact fresh leaf pieces from the adaxial or abaxial sides to a deep of about 100 μm, or intact cells by imaging thick cross sections. In some cases, CLSM can detect the autofluorescence of phenolic compounds as shown in Fig. 1a–c for the anthocyanins in a cherry plum (*Prunus cerasifera*) leaf. The superficial view indicates the presence of anthocyanins in the leaf cuticle (Fig. 1a). Moving inside the leaf, anthocyanins in the epidermal cell vacuoles appear (Fig. 1b) and further deeper the palisade cells are evidenced with chloroplast fluorescence (green) located in the cytoplasm near the cell wall and anthocyanin fluorescence (magenta) in the vacuoles (Fig. 1c).

In red basil (*Ocimum basilicum* cv. Red Rubin), the co-localization of caffeic acid derivatives (namely, caftaric, chicoric, and rosmarinic acids), and anthocyanins (cyanidins) was evidenced in the adaxial epidermal cells (Fig. 1d, e). Identification of the class of compounds can be confirmed by fluorescence micro-spectrometry (Fig. 1f). Under 405 nm excitation, two emission bands peaked at about 485 nm and

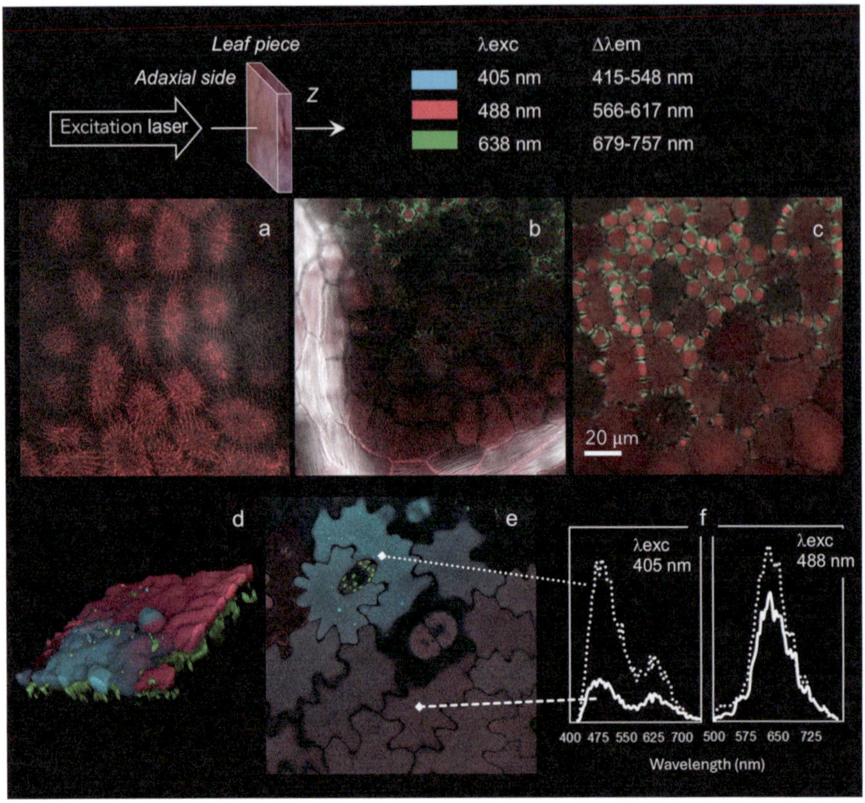

Fig. 1 Image acquisition of anthocyanin autofluorescence by confocal laser scanning microscopy of a cherry plum (*Prunus cerasifera*) leaf piece at different depths (0 μm (**a**), 15 μm (**b**), and 30 μm (**c**)) from the adaxial side. Three-dimensional (**d**) and 2-D (**e**) view of a red basil (*Ocimum basilicum* cv. Red Rubin) leaf adaxial surface. Light-blue emission corresponding to caffeic acid derivatives was acquired in the 415–548 nm range under λ_{exc} at 405 nm. Magenta emission (548–622 nm) under $\lambda_{exc} = 488$ nm is due to anthocyanins. Green emission (672–765 nm) under λ_{exc} at 405 nm comes from chlorophyll in the mesophyll chloroplasts. Fluorescence spectra of single epidermis cells under 405 and 488 nm excitation wavelengths (**f**)

630 nm assigned to caffeic acids and anthocyanins, respectively, are visible. Under the 488 nm excitation, only the anthocyanin fluorescence band is detected.

In the liquid state, at neutral pH, flavonoids do not possess significant autofluorescence quantum yields. Their localization in the leaf tissues by CLSM is therefore possible pending the use of a flavonoid fluorescence inducer. For this purpose, Diphenyl boric acid 2-amino-ethyl ester (DPBA) has been widely employed in plant fluorescence microscopy studies. However, users of this technique must be aware that the reactivity of DPBA with flavonoids depends on the compound structure (Fig. 2).

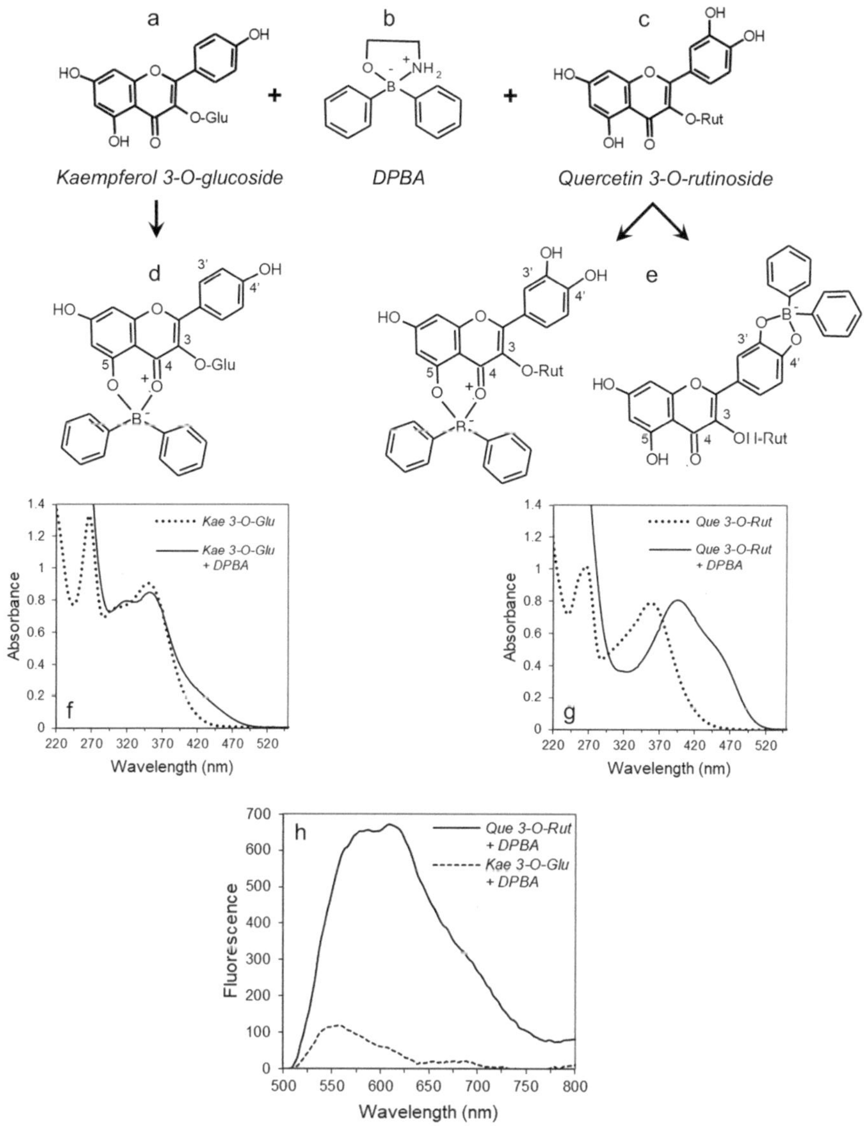

Fig. 2 Complexation of two common plant flavonoids, Kaempferol 3-O-Glucoside (**a**) and Quercetin 3-O-Rutinoside (**c**), representative of monohydroxy and ortho-dihydroxy B-ring flavonoid glycosides, respectively, with Diphenyl boric acid 2-amino-ethyl ester, DPBA (**b**). Binding of (**a**) to (**b**) can occur only at the 5-hydroxy-4-keto (**d**); while reaction of (**c**) with (**b**) results in complexation at both the 3′,4′-o-diphenolic and the 5-hydroxy-4-keto sites (**e**). Absorption spectra in phosphate saline buffer (pH 6.8) of 40 μM Kae 3-O-Glu (**f**) and Que 3-O-Rut (**g**) without and with the addition of DPBA (DPBA/flavonoid molar ratio = 5). Fluorescence spectra of the DPBA-flavonoid adducts (**d** and **e**) uder eccitation at 488 nm (**h**)

For example, in the case of the B-ring ortho-dihydroxylated Quercetin 3-O-Rutinoside (Que 3-O-Rut), DPBA complexation can occur at both the 3′,4′-o-diphenolic and the 5-hydroxy-4-keto competing chelating sites (Fig. 2e), while the B-ring monohydroxy Kaempferol 3-O-Glucoside (Kae 3-O-Glu) can bound DPBA only at the 5-hydroxy-4-keto group (Fig. 2d).

Furthermore, DPBA complexation to Que 3-O-Rut induces a large shift (about 40 nm) to longer wavelengths in the absorption spectrum and the appearance of a second absorption band at around 460 nm (Fig. 2g) (the contribution of the longer wavelength blue absorption band relative to the shorter wavelength one increases by increasing the DPBA/flavonoid molar ratio [25]). In the DPBA–Kae 3-O-Glu complex, only a small increase in absorbance around 440 nm is observed (Fig. 2f). The blue-excited fluorescence of the DPBA-flavonoid adducts resulted to be much higher for Que 3-O-Rut than Kae 3-O-Glu (Fig. 2h).

Therefore, in studying the plant tissue localization of flavonoids by employing DPBA, attention must be paid to the limit of the technique in order to avoid erroneous conclusions. The correct protocol to be used should be based on the following knowledge:

(1) The polyphenolic composition of the species under investigation. (2) The in vitro efficiency and specificity of DPBA as fluorescence inducer for the main different polyphenol compounds to be detected. (3) Not all the subclasses of flavonoids can be stained by DPBA. (4) Glycosylation of the molecules, especially at the C3 position as commonly found in nature, can limit DPBA complexation. (5) In vitro absorbance and fluorescence spectral properties of the DPBA-phenolic complexes. (6) Behind flavonoids, even hydroxycinnamic acids possessing dihydroxy groups, such as caffeic acid derivatives, can react with DPBA changing their absorption spectrum and fluorescence properties. Choose the right combination of excitation/detection wavelengths to discriminate the fluorescence contribution of different compounds. (7) Flavonoids bound to cell walls may have a different reactivity to DPBA than those in a free form into the cell vacuoles.

An example of CLSM fluorescence imaging for the leaf tissue localization of flavonoids in *Phillyrea latifolia* is reported in Fig. 3. Sun and shade leaf cross sections were treated for 2 min by 0.1% DPBA in phosphate saline buffer (pH 6.8) and then observed by the CLSM under excitation at 488 nm and detection in the 570–640 nm waveband. The vacuolar localization of flavonoids in the first layer of palisade and in some adaxial epidermis cells of the sun leaf is clearly evident (Fig. 3a). In the shade leaf, less intense yellow fluorescence was observed only in the cuticle as well as in the outer periclinal and anticlinal walls of the adaxial and abaxial epidermis cells (Fig. 3b).

The destructive analysis of *P. latifolia* leaf extracts showed that sun leaves contained much higher concentrations of the B-ring ortho-dihydroxylated quercetin and luteolin glycosides with respect to shade leaves [26]. On the other hand, the B-ring monohydroxy apigenin glycosides were present at the same level in the leaves under both light regimes. Wall-bound flavonoids were similar in sun and shade leaves and consisted mainly in apigenin glycosides [26].

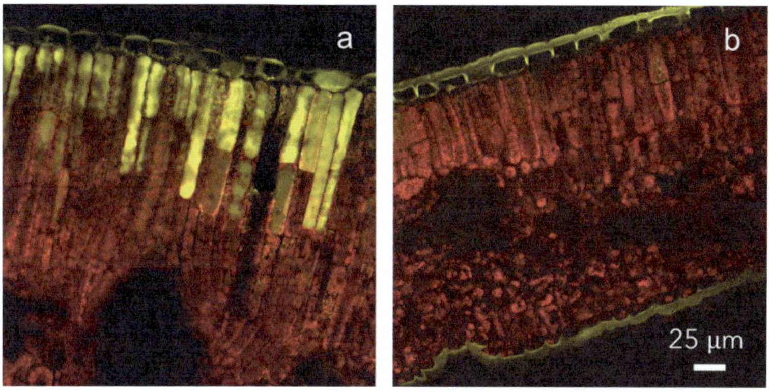

Fig. 3 Merged CLSM fluorescence images of sun (**a**) and shade (**b**) *P. latifolia* leaf cross sections stained with 0.1% DPBA in phosphate saline buffer (pH 6.8) acquired in the 570–640 nm waveband at λ_{exc} of 488 nm and 718–795 nm at λ_{exc} of 638 nm for flavonoids (yellow) and chlorophyll (red), respectively

According to these data and to the in vitro reactivity to DPBA and related fluorescence properties of compounds as explained above, the intense yellow fluorescence observed in the mesophyll of sun leaves could be assigned to quercetin and luteolin glycosides. The free form of apigenin glycosides in the vacuoles could not be detected, while these compounds could contribute to the fluorescence signal observed in the cuticle and epidermis cell walls.

5 Plant Disease-Related Studies

Early non-destructive detection of pathogen attacks to plants can be useful to perform precision phytosanitary treatments and then to prevent yield loss and economic damage to the food industry. It would also contribute to the reduction of environmental pollution by excess use of pesticides. Fluorescence spectroscopy can surely contribute to develop efficient plant disease diagnostic tools passing through the identification of the suitable target molecules and the understanding of mechanisms involved in the plant-pathogen interaction.

The uptake and translocation of a fluorescence-marked penicillin in citrus plants was monitored by fluorescence microscopy, as a model to study the movement of antibiotics through the plant vascular system [27]. This application is significant to better understand the efficacy of antibiotics in controlling plant diseases and in view of a more precise and environmentally safe use of agrochemicals.

Plant activators have a fundamental role in inducing the systemic resistance of plants to pathogen and insect attacks, in an environmentally friendly manner. In order to get more insights on their mechanisms of action, fluorescent-labelled activators were developed and localized into the leaf cells by fluorescence microscopy [28]. The method also allowed to study the interaction of the activators with key proteins of the salicylic acid or the jasmonic acid-ethylene pathways of the plant immune system activation. The fluorescence method is more advantageous than the standard methods for mechanism studies based on immunoassays and immunocytochemical analysis, being less expensive, faster, and non-destructive.

The analysis of fluorescence excitation spectra in plant science is usually much less common than that of the emission spectra. It can instead provide useful additional tools of investigation as, for example, non-destructive plant disease diagnostic methods. Such a case is represented by the Dorokhov et al. (2022) work showing the possibility to discriminate healthy seeds of cereals from those infected by the *Fusarium* fungi by the spectral characterization of the fluorescence excitation in the 180–700 nm range [29]. The short waveband between 220 and 450 nm of excitation produced the largest difference between healthy and infected seeds.

Coumarins possess significant fluorescence emission quantum yields in the blue spectral range when excited by UVA radiation. Therefore, even natural coumarins synthetized by certain species can be detected in vivo on intact leaves. The main role of coumarins in plants concerns their antimicrobial properties, as proved in vitro and supported by the large accumulation of these secondary metabolites in response to the plant infection with bacterial or fungi pathogens.

In the tobacco plant-pathogen interaction, the scopolin and scopoletin coumarins are involved, the second one possessing much higher antimicrobial properties than the first. In vivo fluorescence spectroscopy on bacterial infected tobacco leaves (Fig. 4a) showed a significant increase in blue emission with time after infection [30]. Interestingly, the shape of the fluorescence emission spectrum does not change, while that of the excitation spectrum shifts from shorter to longer wavelengths with time (Fig. 4b). This effect was explained as due to the intra-cell movement of scopolin from the vacuole to the cytoplasm and its conversion to the more antioxidant scopoletin. The higher pH present in the cytoplasm with respect to the vacuole (about 7.5 versus 5.5) induces a change in the fluorescence excitation spectrum of scopoletin, caused by the deprotonation at the available hydroxy group (Fig. 4d) that was consistent with the evolution of the leaf fluorescence excitation spectrum with the progress of the infection. On the other hand, scopolin fluorescence is independent of pH due to the glucoside substitution of the hydroxy group in position 7 (Fig. 4c).

Confocal laser scanning microscopy with excitation at 405 nm confirmed the above hypothesis, showing blue fluorescence localized in the vacuole and a more intense emission as lenticels in the cytoplasm, around the chloroplasts, in tobacco leaves infiltrated with a *Pseudomonas syringae* pv. *tomato* suspension (Fig. 4e). No such structures were evidenced in the control leaves infiltrated with sterile physiological solution (Fig. 4f).

Downy mildew is a severe disease that affects grapevine following the infection by the *Plasmopara viticola* fungus. The plants react to the *P. viticola* attack producing the stilbene class of phytoalexins, having an antimicrobial activity such as zoospore mobility reduction and mycelium growth inhibition. Stilbenoids possess a significant fluorescent in the violet-blue spectral region around 400 nm when

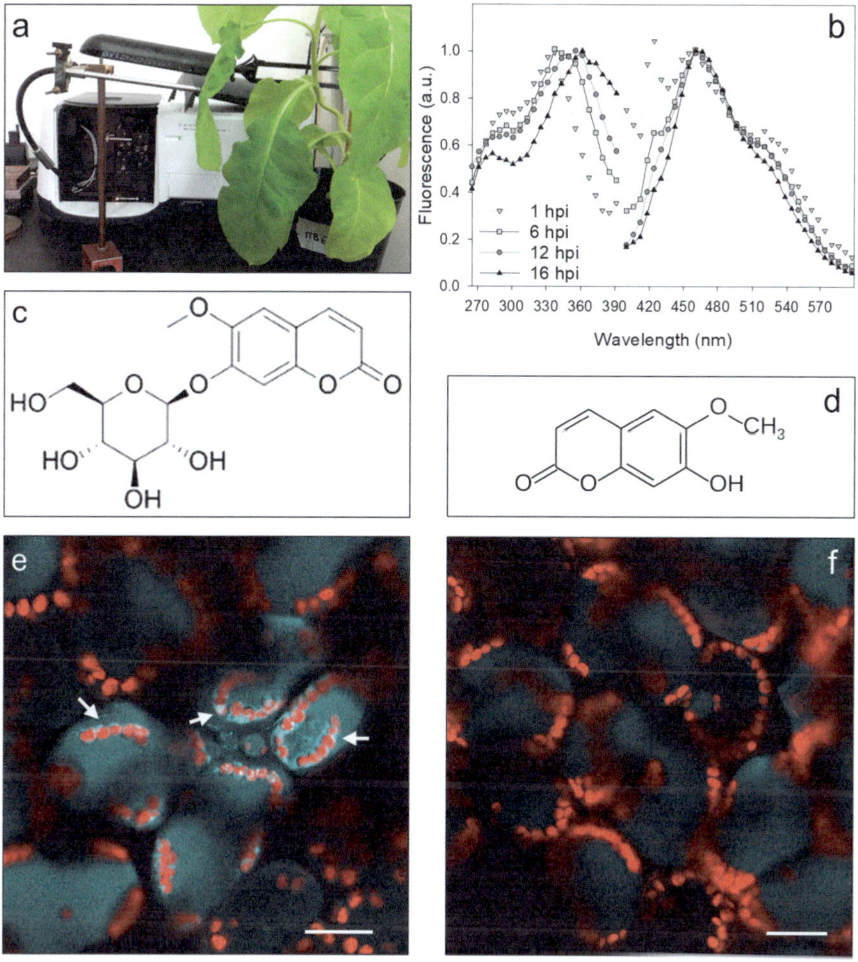

Fig. 4 In vivo fluorescence spectroscopy by using a fibre optics bundle equipped spectrofluorimeter on a *Nicotiana tabacum* L. plant (**a**). Emission and excitation fluorescence spectra of tobacco leaf disks infiltrated with *P. syringae* pv. *tomato* recorded on the adaxial side at different hours post-infiltration (hpi), normalized to the maximum (**b**). The emission spectra were acquired under excitation at 370 nm. The excitation spectra were acquired with emission at 460 nm. Structure of the main phytoalexins in tobacco plants: scopolin (**c**) and scopoletin (**d**). CLSM fluorescence imaging of leaf disks, observed from the adaxial side, acquired at 48 h post-infiltration with *P. syringae* pv. *tomato* (**e**) or with sterile physiological solution (**f**). Light-blue fluorescence, under excitation at 405 nm and detection at 430–498 nm, represents localization of coumarins. Red fluorescence coming from the chlorophyll of the chloroplasts was acquired under excitation at 638 nm and detection at 695–765 nm. Arrows indicate the cytoplasmatic localization of blue fluorescence. Bar = 20 μm

excited at 330 nm, even in in vivo leaves. Therefore, they can be considered natural fluorescence probes of the grapevine leaf *P. viticola* infection [31].

In vivo localization of stilbenes in grapevine leaf tissues at different time post inoculation was determined by CLSM and micro-spectrofluorimetry detection of their autofluorescence and used to get more insights on the host–pathogen interaction mechanisms in both resistant and susceptible vine genotypes [32].

The possibility to non-destructively detect and follow the naturally occurring downy mildew infections on leaves directly in a vineyard was proved by using a portable UV-LED-based fluorimeter able to measure the stilbene fluorescence as proximal sensing [33]. Further technological development of this approach could allow for a large-scale temporal and spatial monitoring of the disease, useful for an efficient prevention and precision phytochemical treatments.

6 Fluorescence Probes for In Vivo Plant Stress Detection

The 4-hydroxyphenylpyruvate dioxygenase (HPPD) is an essential plant enzyme being related to the synthesis of the antioxidant tocopherol, significant in protecting the plants under stressful conditions. Indeed, it has been observed that HPPD was over-expressed under different abiotic stress; therefore, its detection could be used as an early plant stress indicator. For this reason, fluorescent probes able to detect HPPD in vivo have been developed. One approach was based on the conformational change occurring in the probe upon binding to HPPD that leads to a significant fluorophore emission with respect to a quenched state of the folded probe form before binding [34]. The chosen Nile blue fluorophore was not the ideal being its emission overlapped to ChlF. Fluorescence images of *A. thaliana* and *Brassica napus* L. seedlings under different stress conditions were acquired by a multispectral laser imager, with excitation at 635 nm and emission at 670 nm, finding a good relationship between the fluorescence intensity and the plant level of HPPD.

A different fluorescent probe worked on an excimer-monomer equilibrium mechanism [35, 36]. In the aggregated state, the probe blue fluorescence was absent; the disaggregation of the probe upon interaction with HPPD determined the increase of blue fluorescence proportional to the HPPD concentration. Fluorescence images of *A. thaliana* seedlings were acquired with excitation at 365 nm and emission at 400 nm. The choice of the blue-emitting pyrene as a fluorophore and the related acquisition configuration has the advantage to avoid interference from chlorophyll fluorescence in the red spectral region.

Flubactin represents another fluorescent probe, designed by the same above research group, able to visualize in vivo the protein expression level of the ABA (abscisic acid) receptor PYR1, related to the environmental stress condition of plants. Flubactin fluorescence was proved to be an efficient early index of salt stress in *A. thaliana* seedlings [37].

Detection of isoprene in vivo was performed by using a coumarin fluorescence probe [38]. The method was initially developed for an in vitro detection of air samples emitted by plants, dissolved in a cuvette containing the probe in solution and spectrofluorimetric measuring the resulting fluorescence with excitation at 397 nm and emission at 496 nm, with a limit of detection of 1.6 ppm. Isoprene fluorescence micro-imaging of plant leaves under different kind of stress was also showed to be possible.

The potential of a new fluorescent probe, made of a synthesized flavonoid embedded in a commercial surfactant, for the detection of the benfluralin herbicide in plants was reported [39]. Root segments of *A. thaliana* and *Ricinus communis* L. (castor) seedlings incubated with the fluorescent probe were micro-imaged under excitation in the 340–380 nm range and emission above 425 nm. After the seedlings were incubated with the herbicide, a strong quenching of the probe fluorescence was observed. Following this approach, the technique could be further developed and optimized in order to monitor by fluorescence other pesticides in living plants, and important to support an herbicide biosafety risk assessment or in herbicide-plant interaction studies.

Pang et al. (2024) reported a broad overview of 25 recently developed fluorescence probes emitting in the NIR spectral region applied to in vivo plant imaging of cell organelles, the uptake and trafficking of nanoparticles in plant tissues, reactive species, metal ions, plant hormones (auxin), signalling molecules (NO), and injurious substances (NaCl, hydrazine, formaldehyde) [40]. The NIR emission of these probes in the 650–900 nm range is certainly an advantage with respect to visible-emitting probes for imaging more deeply located compounds. However, in some cases significant interference by chlorophyll fluorescence can occur, suggesting the need of a next-generation probes with emission further shifted to longer wavelengths.

7 Micronutrient Detection

X-ray fluorescence spectroscopy (XRF) represents a powerful non-destructive analytical tool for the in vivo determination of the elemental composition of plants. Non-destructive detection of macro- and micronutrients represents a useful tool to test the healthy status of plants and as complementary technique in ecological and physiological studies. XRF is one of the advanced spectroscopic techniques to identify low levels of element deficiency by analyzing the plant materials directly. It has the advantage of being ecofriendly, while conventional destructive methods use chemicals and produce harmful residues.

The kinetics of nutrient uptake and the distribution of elements in different leaf tissues was proved to be detectable [41]. However, possible side effects on plants due to the X-ray radiation exposure must be considered and avoided by choosing the proper protocols of measurements [42].

A comprehensive description of the different XRF techniques and instruments available as well as their applications in plant science is reported in the review by Singh et al. (2022) [43].

Interesting investigation by synchrotron-based X-ray fluorescence concerned the penetration and localization in plant tissues of the vital zinc micronutrient in the wheat [44] and apple [45] species.

The availability of handheld portable XRF spectrometers can favour the diffusion of this technique for in vivo and in-field applications [46, 47].

8 Concluding Notes

In this contribution some recent examples of fluorescence spectroscopy and imaging applied to plant science studies are reported. For forest/agriculture application, the possibility to develop non-destructive fluorescence indices for the early diagnosis of plant diseases or to select cultivars resistant to virulent pathogens appears particularly important. Portable fluorescence sensors can be used directly in the field (or in greenhouses) for a timely detection of abiotic stress plant statuses, as decision support to crop treatment in order to avoid commodity's yield losses. For this aim, clearly, the prospect of using SIF indices retrieved by the plant canopies reflected solar radiation appears of great interest.

The recent development of new fluorescence probes usable in vivo to mark specific indicators of plant stress status is also highly valuable. However, to minimize interference from plant pigment reabsorption or overlapping with chlorophyll fluorescence, the probe emission should be shifted to longer wavelengths into the second NIR region (above 1000 nm).

The use of confocal fluorescence microscopy is fundamental to understand specific processes in plant science and then to develop fluorescence method for macro applications. This was the case to validate the application of the ChlF excitation screening method to indirectly estimate the level of polyphenolic compounds in the leaf epidermis and fruit skin.

I hope that young readers of this chapter will be stimulated to carry on the application of any of these amazing fluorescence techniques in their research studies. I hope that each time they use this approach, they will be enchanted by plant fluorescence, as I have been over the past 40 years.

Acknowledgments My deepest thanks go to my colleagues: Zoran G. Cerovic (Université Paris-Sud-CNRS, Orsay, France), Lucia Guidi (University of Pisa, Italy), Paolo Matteini (National Research Council, Florence, Italy), Patrizia Pinelli (University of Florence, Italy), Massimiliano Tattini (National Research Council, Florence, Italy), Stefania Tegli (University of Florence, Italy), who endured and supported me during about 20 years of fruitful collaborations, as well as to all the younger fellows (too long to list) I met because of them.

References

1. Valcke R (2021) Can chlorophyll fluorescence imaging make the invisible visible? Photosynthetica 59:381–398. https://doi.org/10.32615/ps.2021.017
2. Swoczyna T, Kalaji HM, Bussotti F et al (2022) Environmental stress – what can we learn from chlorophyll a fluorescence analysis in woody plants? A review. Front Plant Sci 13:1048582. https://doi.org/10.3389/fpls.2022.1048582
3. Pérez-Bueno ML, Pineda M, Barón M (2019) Phenotyping plant responses to biotic stress by chlorophyll fluorescence imaging. Front Plant Sci 10:1135. https://doi.org/10.3389/fpls.2019.01135
4. Buschmann C (2007) Variability and application of the chlorophyll fluorescence emission ratio red/far-red of leaves. Photosynth Res 92:261–271. https://doi.org/10.1007/s11120-007-9187-8
5. Tubuxin B, Rahimzadeh-Bajgiran P, Ginnan Y et al (2015) Estimating chlorophyll content and photochemical yield of photosystem II (Φ PSII) using solar-induced chlorophyll fluorescence measurements at different growing stages of attached leaves. J Exp Bot 66:5595–5603. https://doi.org/10.1093/jxb/erv272
6. Escamez S, Terryn C, Gandla ML et al (2021) Fluorescence lifetime imaging as an in situ and label-free readout for the chemical composition of lignin. ACS Sustain Chem Eng 9:17381–17392. https://doi.org/10.1021/acssuschemeng.1c06780
7. Zheng K, Lin H, Hong X et al (2023) Development of a multispectral fluorescence LiDAR for point cloud segmentation of plants. Opt Express 31:18613–18629
8. Bandopadhyay R, Rastogi A, Juszczak R (2020) Review of top-of-canopy Sun-induced fluorescence (SIF) studies from ground, UAV, airborne to spaceborne observations. Sensors 20:1144
9. Agati G, Bilger W, Cerovic ZG (2020) Fluorescence tools for sensing of quality-related phytochemicals in fruits and vegetables. In: Kuswandi B, Siddiqui MW (eds) Sensor-based quality assessment systems for fruits and vegetables. Apple Academic Press, Palm Bay, pp 79–109
10. Bånkestad D, Wik T (2016) Growth tracking of basil by proximal remote sensing of chlorophyll fluorescence in growth chamber and greenhouse environments. Comput Electron Agric 128:77–86. https://doi.org/10.1016/j.compag.2016.08.004
11. Gameiro C, Utkin AB, Cartaxana P et al (2016) The use of laser induced chlorophyll fluorescence (LIF) as a fast and non-destructive method to investigate water deficit in Arabidopsis. Agric Water Manag 164:127–136. https://doi.org/10.1016/j.agwat.2015.09.008
12. Lichtenthaler HK (2021) Multi-colour fluorescence imaging of photosynthetic activity and plant stress. Photosynthetica 59:4–20. https://doi.org/10.32615/ps.2021.020
13. Pineda M, Pérez-Bueno ML, Paredes V, Barón M (2017) Use of multicolour fluorescence imaging for diagnosis of bacterial and fungal infection on zucchini by implementing machine learning. Funct Plant Biol 44:563–572
14. Yao J, Sun D, Cen H et al (2018) Phenotyping of Arabidopsis drought stress response using kinetic chlorophyll fluorescence and multicolor fluorescence imaging. Front Nutr 9:603. https://doi.org/10.3389/fpls.2018.00603
15. Tian Y, Xie L, Wu M et al (2021) Multicolor fluorescence imaging for the early detection of salt stress in Arabidopsis. Agronomy 11:2577
16. Padilla FM, Gallardo M, Peña-Fleitas MT et al (2018) Proximal optical sensors for nitrogen management of vegetable crops: a review. Sensors 18:2083. https://doi.org/10.3390/s18072083
17. Rossi F, Cavigli L, Tuccio L et al (2024) In situ plant peel spectral attenuation of chlorophyll fluorescence excitation as a tool in precision agriculture and photobiology studies. In: Heinemann D, Polder G (eds) Photonic technologies in plant and agricultural science. SPIE, Bellingham, pp 1–7
18. Agati G, Cerovic ZG, Pinelli P, Tattini M (2011) Light-induced accumulation of ortho-dihydroxylated flavonoids as non-destructively monitored by chlorophyll fluorescence excitation techniques. Environ Exp Bot 73:3–9. https://doi.org/10.1016/j.envexpbot.2010.10.002

19. Cerovic ZG, Ounis A, Cartelat A et al (2002) The use of chlorophyll fluorescence excitation spectra for the non-destructive in situ assessment of UV-absorbing. Plant Cell Environ 25: 1663–1676
20. Pinelli P, Romani A, Fierini E, Agati G (2018) Prediction models for assessing anthocyanins in grape berries by fluorescence sensors: dependence on cultivar, site and growing season. Food Chem 244:213. https://doi.org/10.1016/j.foodchem.2017.10.021
21. Agati G, Soudani K, Tuccio L et al (2018) Management zone delineation for winegrape selective harvesting based on fluorescence-sensor mapping of grape skin anthocyanins. J Agric Food Chem. https://doi.org/10.1021/acs.jafc.8b01326
22. Tuccio L, Massa D, Cacini S et al (2022) Monitoring nitrogen variability in two Mediterranean ornamental shrubs through proximal fluorescence-based sensors at leaf and canopy level. Sci Hortic (Amsterdam) 294:110773
23. Dong R, Miao Y, Wang X et al (2020) Estimating plant nitrogen concentration of maize using a leaf fluorescence sensor across growth stages. Remote Sens 12:1139
24. Neugart S, Tobler MA, Barnes PW (2021) Rapid adjustment in epidermal UV sunscreen: comparison of optical measurement techniques and response to changing solar UV radiation conditions. Physiol Plant 173:725–735. https://doi.org/10.1111/ppl.13517
25. Matteini P, Agati G, Pinelli P, Goti A (2011) Modes of complexation of rutin with the flavonoid reagent diphenylborinic acid 2-aminoethyl ester. Monatshefte fur Chemie 142:885–893. https://doi.org/10.1007/s00706-011-0545-z
26. Agati G, Galardi C, Gravano E et al (2002) Flavonoid distribution in tissues of *Phillyrea latifolia* L. leaves as estimated by microspectrofluorometry and multispectral fluorescence microimaging. Photochem Photobiol 76:350–360. https://doi.org/10.1562/0031-8655(2002)076<0350:FDITOP>2.0.CO;2
27. Killiny N, Gonzalez-Blanco P, Santos-Ortega Y et al (2019) Tracing penicillin movement in citrus plants using fluorescence-labeled penicillin. Antibiotics 8:262
28. Chang K, Mei W, Guo T et al (2021) Active fluorescence plant activator N-FBT: a new tool for the study of defense signaling pathway in plants. ACS Agric Sci Technol 1:572–579. https://doi.org/10.1021/acsagscitech.1c00203
29. Dorokhov A, Moskovskiy M, Belyakov M et al (2022) Detection of *Fusarium* infected seeds of cereal plants by the fluorescence method. PLoS One 17:e0267912. https://doi.org/10.1371/journal.pone.0267912
30. Agati G, Brunetti C, Tuccio L et al (2022) Retrieving the *in vivo* scopoletin fluorescence excitation band allows the non-invasive investigation of the plant—pathogen early events in tobacco leaves. Front Microbiol 13:889878. https://doi.org/10.3389/fmicb.2022.889878
31. Poutaraud A, Latouche G, Martins S et al (2007) Fast and local assessment of stilbene content in grapevine leaf by in vivo fluorometry. J Agric Food Chem 55:4913–4920
32. Bellow S, Latouche G, Brown SC et al (2012) In vivo localization at the cellular level of fluorescence induced by *Plasmopara viticola* in grapevine leaves. J Exp Bot 63:3697–3708. https://doi.org/10.1093/jxb/err313
33. Latouche G, Debord C, Raynal M et al (2015) First detection of the presence of naturally occurring grapevine downy mildew in the field by a fluorescence-based method. Photochem Photobiol Sci 14:1807–1813. https://doi.org/10.1039/c5pp00121h
34. Fu Y-X, Liu S-Y, Guo W-Y et al (2022) In vivo diagnostics of abiotic plant stress responses via in situ real-time fluorescence imaging. Plant Physiol 190:196–201
35. Zeng X, Huang Y, Dong J et al (2022) Design of an HPPD fluorescent probe and visualization of plant responses to abiotic stress. Adv Agrochem 1:73–84. https://doi.org/10.1016/j.aac.2022.08.001
36. Zeng X, Ma X, Dong J et al (2023) A protocol for activated bioorthogonal fluorescence labeling and imaging of 4-hydroxyphenylpyruvate dioxygenase in plants. Angew Chem 62: e202312618. https://doi.org/10.1002/ange.202312618

37. Yang J-F, Chen W-J, Zhou L-M et al (2022) Real-time fluorescence imaging of the abscisic acid receptor allows nondestructive visualization of plant stress. ACS Appl Mater Interfaces 14: 28489–28500. https://doi.org/10.1021/acsami.2c02156
38. Liu Z, Wang M, Wu M et al (2024) A sensitive coumarin fluorescence sensor designed for isoprene detection and imaging research in plants. Biosens Bioelectron 248:115998. https://doi.org/10.1016/j.bios.2024.115998
39. Qin T, Zhao X, Jia T et al (2022) A surfactant-assisted approach enables the fluorescence tracking of benfluralin in plants. Spectrochim Acta Part A Mol Biomol Spectrosc 280:121517. https://doi.org/10.1016/j.saa.2022.121517
40. Pang Y, Lu M, Rha H et al (2024) Lighting up plants with near-infrared fluorescence probes. Sci China Chem 67:774–787
41. Kopittke PM, Punshon T, Paterson DJ et al (2018) Synchrotron-based X-ray fluorescence microscopy as a technique for imaging of elements in plants. Plant Physiol 178:507–523. https://doi.org/10.1104/pp.18.00759
42. Montanha GS, Rodrigues ES, Marques JPR et al (2020) X-ray fluorescence spectroscopy (XRF) applied to plant science : challenges towards in vivo analysis of plants. Metallomics 12:183. https://doi.org/10.1039/C9MT00237E
43. Singh VK, Sharma N, Singh VK (2022) Application of X-ray fluorescence spectrometry in plant science: solutions, threats, and opportunities. X-Ray Spectrom 51:304–327
44. Amaral DC, Brown PH (2022) Foliar application of an inositol-based plant biostimulant boosts zinc accumulation in wheat grains: a μ-X-ray fluorescence case study. Front Plant Sci 13: 837695. https://doi.org/10.3389/fpls.2022.837695
45. Xie R, Zhao J, Lu L et al (2020) Penetration of foliar-applied Zn and its impact on apple plant nutrition status: in vivo evaluation by synchrotron-based X-ray fluorescence microscopy. Hortic Res 7:147. https://doi.org/10.1038/s41438-020-00369-y
46. da Costa VM, de Oliveira Lima GJ, Guilherme LRG et al (2023) Towards direct and eco-friendly analysis of plants using portable X-ray fluorescence spectrometry: a methodological approach. Chemosphere 339:139613. https://doi.org/10.1016/j.chemosphere.2023.139613
47. Costa Junior TG, Nunes LC, Feresin Gomes MH et al (2020) Direct determination of mineral nutrients in soybean leaves under vivo conditions by portable X-ray fluorescence spectroscopy. X-Ray Spectrom 49:274–283. https://doi.org/10.1002/xrs.3111

Perspectives on Chlorophyll Fluorescence

Gabriela Cordon, Mailén Petri, Ivana López Valiño, and María Gabriela Lagorio

Contents

The original version of the chapter has been revised. A correction to this chapter can be found at https://doi.org/10.1007/17361_2025_13

G. Cordon (✉)
CONICET, Universidad de Buenos Aires, Instituto de Investigaciones Fisiológicas y Ecológicas vinculadas a la Agricultura (IFEVA), Facultad de Agronomía, Buenos Aires, Argentina

Universidad de Buenos Aires, Facultad de Agronomía, Área de Educación Agropecuaria y Ambiental, Buenos Aires, Argentina
e-mail: gcordon@agro.uba.ar

M. Petri and M. G. Lagorio
CONICET, Universidad de Buenos Aires, Instituto de Química Física de los Materiales, Medio Ambiente y Energía (INQUIMAE), Facultad de Ciencias Exactas y Naturales, Buenos Aires, Argentina

Universidad de Buenos Aires, Facultad de Ciencias Exactas y Naturales, Dpto. de Química Inorgánica, Analítica y Química Física, Buenos Aires, Argentina

I. López Valiño
Universidad de Buenos Aires, Facultad de Agronomía, Área de Educación Agropecuaria y Ambiental, Buenos Aires, Argentina

CONICET, Universidad de Buenos Aires, Instituto de Química Física de los Materiales, Medio Ambiente y Energía (INQUIMAE), Facultad de Ciencias Exactas y Naturales, Buenos Aires, Argentina

M. G. Lagorio (ed.), Comprehensive Series in Photochemical and Photobiological Sciences, Fluorescence of Living Plants 1, https://doi.org/10.1007/17361_2025_4
Published online: 29 March 2025

Abstract The need to produce food for a growing population in a sustainable way implies thinking of new solutions that include technology. Traditional agricultural practices, which are often highly resource-intensive and generate high environmental impacts, are no longer viable in the long term. In this sense, smart agriculture can provide answers to many of the problems directly related to the sustainable feeding of 10 billion people by 2050. The integration of advanced technologies is transforming agriculture into a more efficient and sustainable model, allowing for large-scale food production without compromising ecological and environmental balance. Sensors that use chlorophyll fluorescence to monitor plant health can make a significant contribution to adjusting agricultural practices accordingly. In this chapter, we introduce advancements in machine learning and artificial intelligence associated with chlorophyll fluorescence technologies. We also present some cutting-edge examples currently available on the market that use chlorophyll fluorescence as a source of information, including sensors for terrestrial and marine applications, and microscopy for variable chlorophyll fluorescence and fluorescence imaging. These innovations offer sustainable solutions for smart agriculture.

Keywords Chlorophyll fluorescence imaging, Microscopy for variable chlorophyll fluorescence, Sensors, Smart agriculture

1 Food Production and Smart Agriculture

It is estimated that by 2050, the world population will reach 10 billion people [1], raising two closely related issues: how to increase food production sustainably while, simultaneously, reducing greenhouse gas emissions (GHGs) from agricultural production and deforestation associated with the conversion of forests into agricultural land. Traditional agricultural practices, which are often highly resource-intensive and generate high environmental impacts, are no longer viable in the long term.

The 2030 Agenda for Sustainable Development includes several sustainable goals linked to technological changes in agricultural production to prevent environmental degradation and improve living conditions [2], emphasizing that agriculture must adapt to meet the growing demand for food while minimizing environmental impact and preserving natural resources for future generations.

In this sense, smart agriculture can provide answers to many of the problems directly related to the sustainable feeding of a growing global population [3], representing the integration of new and advanced technologies—often referred to

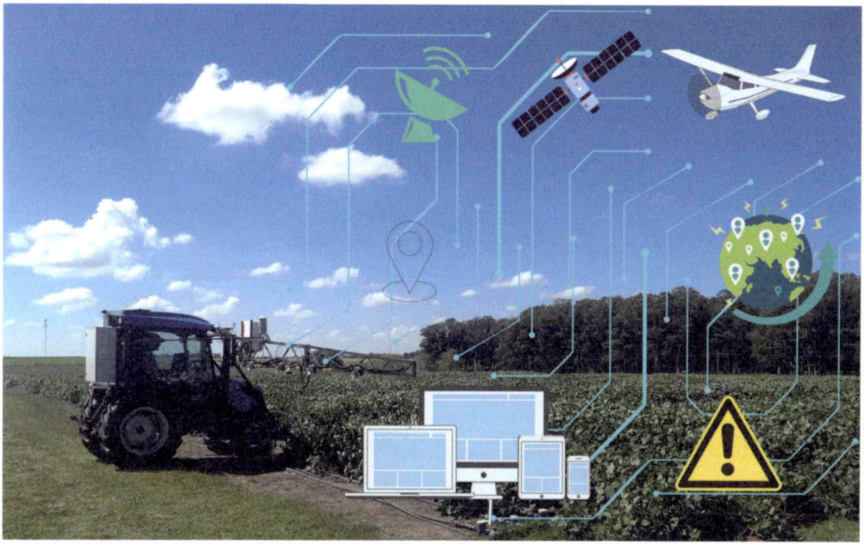

Fig. 1 Pictorial representation of some technologies included in smart agriculture

as digital agriculture or agriculture 4.0 [4]—into a single system to help farmers improve the production and processes of their crops and livestock [5]. According to Kumar and Chouriya [6], smart agriculture integrates artificial intelligence (AI) and Internet of Things (IoT) technologies to address the emerging challenges posed by global population growth, climate change and resource scarcity. This integration is transforming agriculture toward a more efficient and sustainable model, enabling for large-scale food production without compromising ecological and environmental balance.

Clearly, smart agriculture covers a very wide range of technologies ranging from mobile applications for decision-making, field sensors to monitor multiple variables (soil moisture, conductivity, temperature, diseases, plant greenness, presence of weeds), remote sensing techniques for collecting real-time data (and also historical data), which allows for including a great deal of detail of spatial and temporal variability in the information collected, and drones and robots to automate production processes [4] (Fig. 1).

Crop monitoring involves real-time tracking of environmental conditions such as temperature, humidity, and solar radiation, alongside rigorous monitoring of nutrient levels, which can be adjusted through selective fertilization, and the detection of diseases that can be treated by spraying pesticides. Weeds can also be efficiently detected by specific cameras or sensors and subsequently eliminated with targeted herbicide applications [7].

Sensors that use chlorophyll fluorescence to monitor plant health can significantly contribute to adjusting agricultural practices, offering several advantages: their measurements are noninvasive, harmless to plants, and can be repeated without affecting development. These sensors provide high precision and specificity, delivering a direct measure of the physiological state of plants, and their ability to record data in real time ensures continuous monitoring of plant health, enabling rapid responses to detected problems. Techniques based on the emission of light by the chlorophyll-a contained in plants vary in terms of the excitation source, which allows us to differentiate between passive methodologies (sunlight) and active ones that use artificial light to excite chlorophyll-a molecules. Some sensors employ variable fluorescence, recording signals as a function of time for kinetic analysis, while others evaluate the spectral distribution of chlorophyll fluorescence emission, recording signals as a function of wavelength. Additionally, the monitoring scale—whether at the leaf, plant, or vegetation cover level—can vary, providing these sensors with wide versatility.

2 Machine Learning (ML) and Artificial Intelligence (AI)

Machine learning and artificial intelligence methodologies are revolutionizing data analysis and decision-making in many fields. While machine learning uses algorithms and statistical models to enable machines to learn from data and make predictions or classifications without being explicitly programmed for each task, artificial intelligence encompasses a broader range of techniques that allow machines to simulate aspects of human intelligence, such as reasoning and decision-making.

The benefits of these methodologies include the ability to handle large volumes of data, continuously improving performance as more data is accumulated and discovering patterns and relationships that may not be obvious to the naked eye. In addition, they enable complex processes to be automated and tasks to be performed with high accuracy and efficiency. These machine learning and artificial intelligence developments were quickly integrated into chlorophyll fluorescence analysis, and it is anticipated that their impact will continue to grow in the near future.

Chlorophyll fluorescence analysis is an area of study that has been feeding into different disciplines: biology, plant physiology, physical chemistry, biophysics, environmental science, agronomy, food science, aerial and satellite engineering, etc. The enormous amount of data provided by the different methodologies for recording chlorophyll fluorescence makes modern data management necessary, including big data (neural networks, support vector machines, etc.) as a relevant area for the coming years [8–10].

According to a very recent review by Elbasi and colleagues [7], it is possible to implement innovative solutions that not only increase production but also promote the sustainability of the two main areas of agricultural production: crops and livestock, considering the particularities and needs of each of them. To achieve this, Elbasi and collaborators propose incorporating IoT devices, wireless and wired networks, cloud computing, artificial intelligence, and software management systems, among other advances [7].

Statistical models are straightforward and typically require few parameters but face challenges when the relationship between dependent and independent variables is nonlinear. In such cases, ML and deep learning (DL) methods may be more suitable for identifying the relationships between variables, and they have been recently used to estimate crop yield [11–14]. Over the last few years, machine learning techniques have been used to solve various problems in agriculture, especially crop type classification and crop yield estimation [15–19].

To predict wheat yields in the Indo-Gangetic Plains, for example, Liu et al. used machine learning and deep learning methods [20], integrating sun-induced fluorescence (SIF) satellite variables, vegetation indices, and climatic variables, and looked for correlations between variables using different data processing approaches. In particular, they used linear regression methods, machine learning methods, and a deep learning method and found that ML and DL were superior to linear regression methods in predicting wheat yield in the studied plains.

Oliveira et al. applied AI techniques to assess the quality of peanut seeds by combining various variables, including seed area, length, width, coat brightness, chlorophyll fluorescence, anthocyanin content, and reflectance at specific wavelengths, to identify high-quality lots of peanut seeds. They suggested that combining these markers with AI could open the way for automation of postharvest seed quality analysis processes in the future [21].

Lu et al. studied the possibility of detecting chilling damage in cucumbers through the analysis of chlorophyll fluorescence combined with the application of the UMAP-GA-FCM (uniform manifold approximation and projection-gene algorithm-fuzzy c-means) algorithm. They employed principal component analysis to explore the information obtained about chlorophyll fluorescence parameters and evaluate the differences between samples with different degrees of chilling injury. Particularly, through principal components analysis (PCA), they identified those key photosynthetic parameters that were sensitive to chilling injury. Then, they used the UMAP method to extract features from chlorophyll fluorescence parameters that were used as inputs in a classification model [22].

Park et al. studied the effect of abiotic stress in garlic plants by means of chlorophyll fluorescence and multispectral imaging, assessing the effectiveness of stress classification using the ratio of chlorophyll fluorescence combined with a classification model based on the partial least square discrimination analysis (PLS-DA) [23].

A complete review about the use of machine learning methods for assessing photosynthetic activity from local or remote optical measurements (hyperspectral imaging, solar-induced chlorophyll fluorescence, local chlorophyll fluorescence imaging, and fast and delayed chlorophyll fluorescence induction) was published by Khruschev et al. [24].

3 Different Platforms of Chlorophyll Fluorescence Analysis

Considering the four most popular methodologies for chlorophyll fluorescence analysis, pulse amplitude modulated (PAM) fluorometry, fast rise of chlorophyll-a fluorescence (OJIP), fast repetition rate (FRR), and sun-induced fluorescence, it has been observed that PAM fluorometry was the most popular during the decade from 2000 to 2019. In fact, Zavafer et al. conducted a study to determine which analysis platform was the most widely used and found that, over that period, the order of popularity based on the number of exclusive publications was PAM (2459 exclusive publications) > OJIP (1073 exclusive publications) > FRR (421 exclusive publications) > SIF (176 exclusive publications) [25]. The SIF technique, being the most recent, has shown a rapid growth in the number of publications in recent years, and according to these authors, the trend suggests that SIF could become one of the most relevant platforms in the near future, surpassing the other methodologies in the number of publications.

It is important to note that the investment required to use each of these methodologies varies considerably. For instance, OJIP measurements require relatively affordable equipment and provide data that is highly sensitive to biotic and abiotic factors that can affect plants. These characteristics have made OJIP the most widely used methodology in developing countries [25]. At the opposite end of the spectrum in instrumental complexity is airborne- or satellite-based SIF monitoring; due to its high cost, this methodology is mainly limited to countries with greater investment capacities.

As for the FRR technique, it is important to note that it is particularly suitable for use in conditions of low chlorophyll concentrations, such as in the open ocean and in general oligotrophic environments, where the amount of biomass is low [26]. Until another methodology is developed that can be used with low phytoplankton concentrations, this technique will continue to be used, although probably confined to studies in oceans and nutrient-poor water bodies [25].

4 Robotics

While incorporating chlorophyll fluorescence analysis in robotic systems might seem futuristic, this technology is already in development. In fact, in 2015, Nishina, from the Research Center for High-technology Greenhouse Plant Production of Ehime University, developed an intelligent greenhouse system using a chlorophyll fluorescence imaging robot to monitor the growing conditions of tomato crops [27]. He referred to this technology as intelligent greenhouse systems, which have application in commercial tomato production greenhouses. He also stated that it was based on the speaking plant approach (SPA). The SPA is a concept in agricultural technology that refers to the idea of plants communicating their physiological state, needs, and stress levels by means of experimental signals detected by monitoring devices and interpreted by experts. Chlorophyll fluorescence is the primary method employed in this approach. This work was based on previous studies also centered in SPA [28].

At the University of Arizona, Newcomb and Shakoor developed a "robotic field scanner phenotyping platform" for genetic studies of the maximum quantum yield of photosystem II (Fv/fm) by means of a chlorophyll fluorescence imaging system [29].

5 Cutting-Edge Examples Available on the Market Employing Chlorophyll Fluorescence as a Source of Information

5.1 Chlorophyll Fluorescence-Based Weed Detection

There are currently three systems on the market that use technology to detect weeds based on (1) the Normalized Difference Vegetation Index (NDVI) spectral index, derived from the light reflected by the plant cover (reflectance), (2) chlorophyll fluorescence, and (3) artificial vision, with all cases involving local (hardware) processing in real time. It is worth mentioning that the first two are very fast, while the analysis of the images is slower as it considers the morphology and texture of the plant being evaluated. In general, in addition to the sensor that detects the presence of weeds, the systems usually have speed indicators that improve the synchronization between the time the weed is identified and its removal—which can be carried out in different ways—the most widespread being herbicide spraying or boiling water, with herbicide application being the most widespread mechanism for extensive cultivation.

Turning to detection methodologies, NDVI assessments can be made using multispectral cameras, passive or active sensors mounted on the wings of sprayers, with the difference in these cases being spatial resolution and whether it is possible to access the data in real time. According to some specialists, active sensors allow for more effective control, which translates into greater herbicide savings. In addition, they process information quickly, enabling real-time improvements in practices. The Trimble® WeedSeeker is one of the most widely used sensors in the field.

Another system uses chlorophyll fluorescence as a method of detecting weeds, and because the signal emerges directly from the chlorophyll-a molecules present in the reaction centers of the photosystems, it is a very specific signal with no interference from other non-plant material, ensuring that only weeds are detected and sprayed, which significantly reduces the application of chemicals by up to 90%, thus reducing production and environmental costs [30]. This detection principle uses the Rometron® Weed-It sensor [31], and in addition, this system uses artificial light to excite the plants, making it an active system, which is an improvement for working on cloudy days or even at night.

Finally, developments in the field of machine vision raise expectations for its potential use and complimentarily with AI, as this technology allows a machine to "see" through cameras and computers in replacement of human vision, enabling a wide capacity to automate AI systems according to Dhanya et al. [32]. The data collected by the cameras must be processed and analyzed to correctly understand the visual information from the real world to generate numerical or symbolic information that can be used by a computer or AI to facilitate decision-making and task automation. This symbiosis between machine vision, AI, and machine learning or deep learning will improve the discrimination of weeds in the environment and even allow differentiation between plant species, potentially allowing discriminating crops from weeds or distinguishing between different weeds [33].

5.2 *Advanced Equipment for Assessing Quantum Yields and Plant Health*

Another useful piece of equipment in intensive farming practices is based on a fluorescence technique called Laser Induced Fluorescence Transient fitted by Fast Repetition Rate (LIFT-FRR) that uses a sequence of short flashlets ~1 μs of light (450 nm in the monochromatic version but can be adapted to users' requirements) to activate the photosynthetic apparatus and record the resulting changes in fluorescence yield [34].

According to the developer Solience®, this type of methodology could be useful in intensive farming practices as it would, for example, anticipate crop yields in variable weather conditions. Solience® claims that it is possible to make fertilizer or irrigation decisions based on these methodologies applied in real time, allowing early detection of water or nutrient stress before physiological and morphological damage to crops occurs. In this regard, the company Solience® rescues on its official website some extremely interesting concepts from a survey conducted in July 2016 on the state of plant phenotyping carried out by the International Plant Phenotyping Network (IPPN), an association representing the world's leading plant phenotyping centers as stated in the IPPN webpage [35]:

> "Sustainable food security and increasing availability of plant biomass for human nutrition and bioindustries is the key challenge for the coming decades. Understanding the physiological and genetic basis of plant growth and its translation to crop improvements is vital to address the future challenges. This requires concerted action to closely interact between different stakeholders to evaluate and map the demand for phenotyping, available infrastructure, to evaluate the opportunities and limitations and to discuss strategies for the development of the plant phenotyping community."

There are different instrumental options in the configuration of the commercial equipment adapting to the different requirements of both terrestrial and marine environments, with terrestrial LIFT versions offering different excitation wavelengths, such as 450 nm (monochromatic version), 450, 470, 510, and 530 nm (color version). Depending on the excitation source, different remote measurement ranges can be implemented: up to 3 m distance using an light-emitting diode (LED)-based excitation source, up to 5 m in the Light Induced Fluorescence Transient Remotely (LIFT-REM) version with one or four excitation channels with software-controlled wavelength and excitation power selection and up to 50 m distance using a laser light source. In addition, there is a choice of detection system; the LIFT-SPEC-1.0 model is equipped with a spectrometer with a spectral range between 325 and 812 nm, while the LIFT-CAM-1.0 uses a dual camera system for 3D image reconstruction [36].

Finally, it is possible to use different operation protocols to obtain very specific and varied information related to photosynthesis and therefore to plant health: (1) functional absorption cross section of photosystem II (PSII); (2) yield of charge separation in PSII; (3) kinetics of electron transport; (4) probability of energy transfer; (5) size of the oxidized portion of PQ pool; (6) carotenoids quenching; and (7) donor side quenching [36].

5.3 *Space Mission for Reflectance and Chlorophyll Fluorescence Measurements*

The LIFT-FRR-based measurements currently record reflectance spectra using an online spectrometer, but there are plans to upgrade to a two channel spectrometer, which will enable the simultaneous acquisition of both reflectance spectra and solar irradiance spectra, allowing for correlation with data from the upcoming FluorescenceEXplorer (FLEX) satellite mission [37]. The FluorescenceEXplorer (FLEX) mission, to be launched into space in 2025, will provide global fluorescence maps of vegetation, offering insights into the photosynthetic activity and plant health and stress levels. According to the European Space Agency (ESA), these advancements will enhance our understanding of the global carbon cycle and contribute to progress in agricultural management and food security, critical in meeting the demands of a growing global population [37]. In the next version of LIFT-FRR, the two-channel spectrometer could provide ground truth information to validate the FLEX satellite observations and facilitate the correlation of passive fluorescence signals from FLEX with the active fluorescence signals captured by LIFT, further improving the precision of these measurements.

5.4 *Variable Fluorescence in Aquatic Environments*

In terrestrial environments, the LIFT-FRR-based measurements are widely used to assess photosynthetic damage and to evaluate the net primary production of different environments, while in marine environments, they are commonly employed to assess phytoplankton photosynthesis and primary productivity by means of actively induced measures. According to Solience®, fluorescence transient will be processed to quantify a wide range of biophysical characteristics that determine the photosynthetic performance of marine and freshwater algae and photosynthetic bacteria. They offer different configuration options on their commercial equipment according to the specifications requested by users, including combinations of excitation wavelengths between 445 nm, 470 nm, 510 nm, 530 nm, and 590 nm, available in FRR-based or direct current (DC)-based modes [38]. This enables the instrument to discriminate between different phytoplankton groups based on spectral composition of their light harvesting pigments. Additionally, the instruments can be equipped with eight additional excitation channels operating in DC mode, with an oxygen sensor, a temperature controller, and a magnetically coupled mixer.

Oceanic primary production is known to be limited by the availability of nutrients, such as nitrogen and iron, with nutrient shortages leading to decreased efficiency of oceanic photosynthesis. Variable chlorophyll-a fluorescence is the most sensitive, nondestructive signal detectable in the upper ocean and is widely used in marine environments to evaluate primary production and phytoplankton photosynthesis. Using this technique, Gorbunov and Falkowski [39] evaluated phytoplankton photosynthesis under nutrient-limited conditions and developed a new kinetic-based approach to measure electron transport rates (ETRs), directly and in absolute units, as well as to estimate growth rates in phytoplankton. They further studied the efficiency with which solar energy is used in the phytoplankton photosynthetic process, dividing the absorbed energy for photosynthesis into three terms: the quantum yields of photochemistry, fluorescence, and heat [40]. By measuring seagoing variable chlorophyll-a fluorescence and picosecond fluorescence lifetime, it is possible to obtain information about the first two terms, enabling researchers to understand how phytoplankton responds to nutrient supplies on timescales ranging from hours to months and, over longer periods, to changes in climate.

5.5 *Equipment for Determining the Spectral Distribution of Chlorophyll Fluorescence in Plant Covers*

There is currently a piece of equipment on the market that has the necessary facilities to infer the spectral distribution of chlorophyll fluorescence closely related to the information soon to be provided by the FLEX mission. FloX, the Fluorescence Box, is a hyperspectral instrument for observing solar-induced fluorescence (SIF) and reflectance spectra at the same time. For this purpose, it has two spectrometers with different spectral ranges and resolutions. The equipment that measures SIF is called operational performance FLUO and has a spectral range ~650–800 nm, which allows it to cover the two oxygen absorption bands (O_2A and O_2B) to extract the solar-induced chlorophyll fluorescence signal in the red and far-red using the improved Fraunhofer line discriminator (FLDi) method [41] or the spectral fitting methods (SFM) [42]. This instrument also has a spectral resolution of ~0.3 nm (full width at half maximum, FWHM) and a high signal-to-noise ratio (SNR) ~1,000. On the other hand, the optical performance FULL records reflectance measurements in the visible to near infra-red spectral range ~400–950 nm with a spectral resolution of ~1.5 nm (FWHM) and a moderate signal-to-noise ratio (SNR). This device enables short-distance monitoring measurements from 1 m to significant heights such as 10 m (e.g., wheat, corn) [43] or 50 m (e.g., trees) depending on the objective of the study [44, 45].

These three examples of chlorophyll fluorescence–based sensors show the potential for integrating this technology with emerging tools like deep learning, precision agriculture, and drones, highlighting the role such sensors will play in the future of smart agriculture. The fusion of chlorophyll fluorescence with AI-based sensors undoubtedly leads to a more personalized approach to farming where real-time decisions at the individual plant or small plot level will be taken. This approach ultimately connects the use of sensors based on chlorophyll fluorescence with robust agricultural production that requires fewer resources, minimizes input, and therefore becomes more sustainable over time. However, realizing the full potential of this technology will require sustained investment in research and development to maximize its impact on agriculture.

6 Microscopy for Variable Chlorophyll Fluorescence

The spectroscopic properties of plants often differ when recorded over the entire tissue compared to the chloroplast level [46], highlighting the importance of obtaining photosynthetic parameters at the chloroplast level by combining variable fluorescence methodology with microscopy techniques.

In 2009, a method was developed to measure chlorophyll fluorescence parameters of chloroplasts using an inverted light microscope with a Nipkow disk confocal scanner unit [47]. This setup enabled researchers to measure chlorophyll fluorescence transients in Boston fern (*Nephrolepis exaltata*) and board bean (*Vicia faba*) chloroplasts from the abaxial leaf side and reconstruct a 3D image by capturing fluorescence intensity images across 64 focal planes after a saturation pulse. Using this 3D image and appropriate equations, they obtained images of chlorophyll fluorescence yield parameters, such as Non-Photochemical Quenching (NPQ) and ΦPSII. They observed that the mean ΦPSII values of mesophyll, epidermal, and guard cell chloroplasts decreased with increasing actinic light intensity, with the lowest values in guard cell chloroplasts and progressively higher values in epidermal and mesophyll chloroplasts at all light intensities. The usage of a spinning disk was not ideal because of its small field of view, low optical sectioning ability, and its nonuniform excitation intensity on the measurement region [48], making it difficult to obtain the response from a single chloroplast.

Recently, advances have enabled measurement of Kautsky kinetics using a laser scanning confocal microscope, which offers a greater field of view, uniform excitation intensity on the sample, and the usage of less powerful lasers for measurement, reducing the possibility of photothermal damage to the tissues under study. Tseng and Chu [49] adapted a confocal microscope with a low numerical aperture objective to be able to measure fluorescence kinetics with a resolution reaching the micrometer, better than a single chloroplast, and the millisecond. With this equipment, Kautsky curves of a single chloroplast were obtained, although with a significant amount of noise due to the small number of pixels involved in the measurement. Therefore, they obtained an image of the spatial distribution of the quantum efficiency of the photosystem II inside a living leaf of *Brugmansia suaveolens*, where individual chloroplasts could be distinguished.

In this first iteration of the fluorometer, the confocal laser is used as the measuring light, the source of the actinic light and the saturating pulse to achieve the variable fluorescence. Consequently, it was not possible to apply two types of light simultaneously, which limited both the accuracy of the determination and the number of photosynthetic parameters that could be obtained. In a newer development [50], an inverted confocal microscope was equipped with a red-light LED source in front of the microscope objective, which delivered short and intense pulses capable of saturating the photosystem II (PSII) to the entire sample and also provided continuous actinic light. The blue laser of the microscope acted only as the measuring light, enabling the execution of non-photochemical quenching protocols. Furthermore, this newer version allowed the measurement of the variable chlorophyll fluorescence on three dimensions and overtime, facilitating the detection of fluorescence of a single plastid, which was previously difficult due to the movement of plastids inside the cell that often caused them to leave the field of observation during the time required for the development of non-photochemical quenching. The researchers used this setup to conclude that the photosynthetic differences observed between chloronema and caulonema cells of *Physcomitrella patens* cannot be attributed to intrinsic properties of the PSII, since the photosynthetic parameters derived from both types of cells were similar, but that the net amount of photosynthetic machinery, as shown by the total chlorophyll fluorescence emission and by the fraction of the cell occupied by plastids, differs, explaining the difference between these tissues in the global process of carbon fixation. They were also able to identify *Phaeocystis cordata* subpopulations with low photosynthetic performance and photoprotection capacity and obtained more evidence of light being differently channeled through different leaf tissues and of photosynthesis being largely governed by plants' anatomical features.

The detection of Kautsky kinetics at a single cell level was also carried out on a *Synechococcus elongatus* cyanobacteria, which contains chlorophyll-a, using an optical Fabry-Pérot microcavity and a confocal scanning microscope [51]. The microcavity's influence on the cyanobacteria was evaluated, observing that the quantum yield differs between a resonant and off-resonant microcavity, leading them to propose that it has an effect on the energy transfer in the photosystem II in vivo.

In brief, microscopy for variable chlorophyll fluorescence analysis is useful to understand plant physiology and photosynthesis at a cellular or subcellular level. It enables researchers to analyze photosynthetic activity in individual chloroplasts, providing insights into how different regions of a cell or tissue contribute to overall plant health. The fine-scale analysis provided by microscopy helps in detecting early signs of stress that may manifest in specific cells or tissues before affecting the entire plant. Microscopy for variable chlorophyll fluorescence can probably become in future a routine practice for precision plant science.

7 Chlorophyll Fluorescence Imaging

In recent years, variable chlorophyll-a fluorescence imaging has been used extensively to detect inhomogeneous damage on plant leaves in a laboratory setting. For example, chlorophyll fluorescence imaging was used to observe differences in the spatial distribution of damage, measured as changes on OJIP parameters, between cadmium (Cd) toxicity and zinc (Zn) deficiency in soybean (*Glycine max*), a species sensitive to both metal deficiency and toxicity, and *Noccaea ochroleucum*, a relatively metal-resistant plant [52]. Both treatments tended to have a detrimental effect on photosynthesis, but the location of the damages was different. In soybean, Cd toxicity affected photosynthesis all over the leaf, whereas in *N. ochroleucum*, it affected the veins. On the contrary, Zn deficiency always affected the mesophyll more, away from the veins. They also observed different rates of acclimation to actinic light in various cell types.

In another study, chlorophyll fluorescence imaging was used to observe the spatial distribution of damage in *Arabidopsis* plants treated with plant natural compounds, revealing distinct asymmetric effects of different treatments as well as differences between treated young and old leaves. These findings highlight the benefits of measuring the whole plant fluorescence with spatial distribution over using single-point measurements using clamping fluorometers [53].

Arief et al. [54] used strawberry (*Fragaria x ananassa*) plants under heat stress, drought stress, and a combination of both stresses and observed a reduction of the maximal PSII quantum yield (Fv/Fm) measured with a chlorophyll fluorescence imaging system developed by themselves. Legendre et al. [55] evaluated atrazine effects on the chlorophyll fluorescence of *Catharanthus roseus*, *Lactuca sativa,* and *Impatiens hawkeri*, showing that chlorophyll fluorescence imaging can detect herbicide-induced stress before visual damage.

As mentioned before, chlorophyll fluorescence can also be measured on cyanobacteria, and in a study conducted by Colussi et al. [56], imaging chlorophyll fluorescence was used to study the acclimation of *Trichodesmium erythraeum* IMS101 to nonlethal iron limitation, showing a decrease in the quantum yield of PSII and no significant changes on the efficiency of electron flux between PSII and PSI, indicating the ability of this cyanobacteria to cope with the lack of iron to assemble the many iron clusters in PSI.

Chlorophyll fluorescence imaging may also be employed for commercial purposes on fresh fruits and vegetables. Herppich [57] wrote a complete review of its usage as a way to evaluate postharvest decay, the effects of storage conditions, and the consequences of different sanitation treatments, with studies including lemons, arugula, pepper, and apples. Additionally, as explained by Rühle et al. [58] in their experiments with *Arabidopsis*, chlorophyll fluorescence video imaging can also be useful to conduct automatic screening studies to detect photosynthetic phenotypes.

For more in-depth analysis on using chlorophyll fluorescence imaging to detect different types of stress on plants before visual symptoms develop, Valcke [59], Moustaka and Moustakas [60], and Pérez-Bueno et al. [61] wrote reviews of chlorophyll fluorescence imaging as a tool to detect abiotic stresses, such as metal toxicity or deficiency, drought, ozone damage, freezing damage and heath stress, exposure to herbicides or phytotoxins, and biotic stresses, in order to phenotype the plants based on their responses to the stresses.

Oláh et al. [62] measured chlorophyll fluorescence imaging of *Spirodela polyrhiza* exposed to nickel, hexavalent chromium, and sodium chloride and compared the sensitivity of various parameters fitting them to concentration-dependent nonlinear regression models, some including a hormetic component at the lower concentration range, concluding that generally light-adapted fluorescence parameters were more sensitive than the dark-adapted ones. Furthermore, in some cases, fluorescence imaging was combined with deep learning or the training of computer models to develop a method for automatic detection of contamination or stress via fluorescence imaging.

In most studies, researchers test different models to conclude the best one at recognizing a certain type of stress. One study [63] evaluated six convolutional neural network (CNN) models—AlexNet, GoogLeNet, ResNet50, ShuffleNet, SqueezeNet, and MobileNetv2—for salt stress recognition in soybean seedlings based on three types of chlorophyll fluorescence–derived images (maximum quantum efficiency of PSII, the quantum yield of energy dissipation of PSII, and the inhibitory effect of salt on PSII quantum yield), concluding that ResNet50 outperforms the other models in this setup. Another study [64] developed a classification method based on artificial neural network (ANN) using chlorophyll fluorescence imaging on *Schefflera octophylla (Lour.). Harms* caused by three different water treatments (high, middle, and low water content), three different nitrogen treatments (high, middle, and low nitrogen), and nine levels of water-nitrogen coupling treatments, resulting from the combination of different water and nitrogen treatments. They tested three models—support vector machine, radial basis function, and backpropagation—to develop a plant nitrogen and water status classification and concluded that the radial basis function was more reliable for achieving this purpose on the studied plant.

Chlorophyll fluorescence images can also be used in conjunction with other parameters to better develop stress recognition models. For example, Zhang et al. [65] used chlorophyll fluorescence imaging along with hyperspectral imaging to identify rice plants (*Oryza sativa*) under butachlor, quinclorac, cadmium, and copper stress, developing a high-level fusion-based convolutional neural network model that they concluded had a higher detection accuracy than models using a single data source.

As already mentioned, chlorophyll fluorescence imaging can be combined with other techniques to complement the results. For example, Vítek et al. [66] combined chlorophyll-a fluorescence measurement with Raman spectroscopy to evaluate the effects of an herbicide on the pigment composition of *Helianthus annuus* plants. Vítek et al. [67] also evaluated the effect of the herbicide metribuzin, which is a photosystem II inhibitor, on *Chenopodium album* plants. The temporally and spatially resolved cascade of defense responses was studied noninvasively by combining three imaging approaches: Raman spectroscopy as the principal method, corroborated by chlorophyll-a fluorescence and infrared thermal imaging. They concluded that the combination of Raman spectroscopy, chlorophyll-a fluorescence, and thermal imaging allowed them to monitor the cascade of damage and defense responses following the application of the PSII inhibiting herbicide with high spatial resolution. It also was useful for monitoring numerous abiotic and biotic stresses showing a similar cascade of response.

In some cases, methods of nondestructive diagnosis using characteristics of the photos of the chlorophyll fluorescence parameters were developed. For example, Long et al. [68] studied a method for identification of drought stress in tomato seedlings using chlorophyll fluorescence imaging. They applied three feature optimization algorithms (successive projections algorithm, iteratively retaining informative variables, and variable iterative space shrinkage approach) to select the fluorescence parameters with better correlation. Histogram features and texture features of the images of the chosen parameters were input into three models, linear discriminant analysis, support vector machines, and k-nearest neighbor, to develop one able to identify drought stress classes, concluding that the linear discriminant analysis model had the highest recognition accuracy rate (86.6%). Similarly, Dong et al. [69] developed a method for detecting chilling injury of tomato seedlings based on four chlorophyll fluorescence image features: the fluorescence parameters themselves, histograms, textures, and color coordinates of the images. The eligible parameters were selected using Pearson correlations and were input into a back propagation neural network model to develop the previously mentioned method.

All the previously mentioned examples were carried out at laboratory settings. Nowadays, field experiments are possible using smaller fluorometers that allow measuring of a single plant, where boxes have to be used in order to dark-adapt individual plants. Li et al. [70] carried out both a greenhouse experiment and a field experiment on soybean using multiple herbicide combinations selected according to the local practice of farmers of the area where the study was carried out and used the recommended dosages or half the recommended dosages. They employed a mobile fluorescence sensor, the WEED-PAM® sensor, to measure chlorophyll fluorescence and obtain maximal PSII quantum yield images of the tested plants. In both cases, dark adaptation cover boxes had to be used, and not all plants were tested, but randomly selected ones of each treatment group and of the control group. When using the recommended dosage, the herbicide combinations reduced the Fv/Fm value of the soybean plants in all treatments in the greenhouse and in four of the five field trials. Therefore, the researchers concluded that the WEED-PAM® sensor allows for the quantification of soybean plants' response to the herbicide treatments and can help select the most tolerant cultivars for future harvests. Wang et al. [71] conducted three field experiments on *Alopecurus myosuroides* and used the previously mentioned WEED-PAM® sensor for chlorophyll fluorescence imaging. They used treatments with combinations of herbicides that were acetolactate synthase (ALS) or acetyl CoA carboxylase (ACCase) inhibitors at recommended dosages. In the first field experiment, the seeds used were of a sensitive *A. myosuroides* population, in the second experiment they worked in a field heavily infested with *A. myosuroides* that were reported to be resistant to ALS- or ACCase-inhibiting herbicides, and in the third experiment, they used *A. myosuroides* populations that proved to be resistant to either ALS-, ACCase-, or PSII-inhibiting herbicides. They used the Fv/Fm values to compare sensitivity to the different herbicides, and the plants with known resistance were used as a test for the method, and in general, the results were correct, with resistant populations having equal Fv/fm values to untreated populations and sensitive populations having significantly lower Fv/Fm values. Similarly, to the previously mentioned study, dark adaptation cover boxes had to be used, and not all plants were tested.

Finally, in another paper, Wang et al. [72] also conducted greenhouse and field experiments in which they measured images of Fv/Fm of herbicide-sensitive and herbicide-resistant *A. myosuroides* plants using the WEED-PAM® sensor. When treated with herbicides in their recommended dose, the resistant populations showed Fv/Fm values equal to control groups, whereas sensitive populations had lower Fv/Fm, allowing the differentiation of the two populations even after one day of treatment. Therefore, in their experiments, similarly to the previous study, they concluded that they were able to detect herbicide-resistant weed populations soon after exposure using a chlorophyll fluorescence–imaging sensor. In the mentioned field experiments using small fluorometers, not all plants of the experiment were evaluated, but randomized ones that were assumed to be representative of the whole plot. More advanced equipment, such as the TERRA-REF field scanalyzer, have the fluorometer mounted on a tractor, which allows the measurements to be automatic and on a larger number of plants, but even this equipment cannot be used to measure from significant heights.

Herritt et al. [73] aimed to validate a field-deployed fluorescence imaging system on the TERRA-REF field scanalyzer comparing its results with a commercial handheld fluorometer. To achieve this, they treated *Sorghum bicolor* with 3-(3,4-dichlorophenyl)-1,1-dimethylurea (DCMU), an inhibitor of PSII. They concluded that Fv/Fm showed the maximum correlation between the two types of measuring in both treated and untreated plants. They also saw that the data from the imaging system had less variability and was faster to measure. The study concluded that fluorescence values obtained from the full canopy of a plot in a field appeared to be less variable than obtaining multiple point measurements with a handheld device over the canopy. However, the fluorescence images obtained with this equipment are difficult to analyze, as canopy and soil background can be contained in the same pixel due to image resolution. In a more recent paper, the authors described software developed to automatically process the fluorescence images and calculate the derived fluorescence parameters, such as Fv/Fm. This software is specifically for the TERRA-REF chlorophyll fluorescence imaging system, but the study describes the pipeline used to develop it, which is also applicable to other chlorophyll fluorescence imaging systems [74].

As an objective for the future, it will be interesting to obtain fluorescence imaging measurements from longer distances in order to evaluate a whole field at the same time, saving time and resources. It is necessary to consider that the application of these types of equipment has two limitations: they cannot be used on windy days, and if one wants to use active techniques, the experiments have to be carried out at

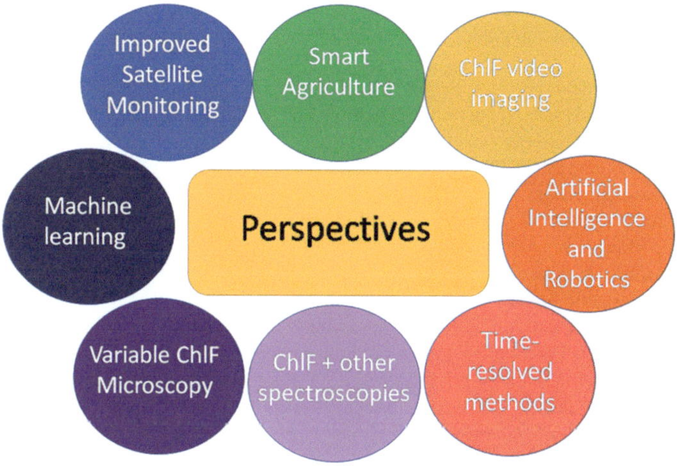

Fig. 2 Pictorial summary of the concepts discussed in this chapter on the future perspectives of plant fluorescence

night or using a setup to cover the sun in order to dark-adapt the leaves. Because these techniques allow a more representative and quicker measurement, these limitations seem worth having.

In conclusion, the integration of cutting-edge technologies such as artificial intelligence, machine learning, robotics, imaging techniques, video imaging, remote sensing, and advanced microscopy has the potential to highly improve the field of chlorophyll fluorescence (Fig. 2). These methodologies, combined with other spectroscopic methods, will further enhance our ability to monitor and analyze plant health and photosynthetic processes with unprecedented precision. These innovations not only promise to deepen our understanding but also offer sustainable solutions for agriculture, environmental monitoring, and beyond. The concepts presented in this chapter represent just a small sample of the numerous possibilities for the future of plant fluorescence analysis which will be surely enriched by multidisciplinary approaches.

References

1. United Nations (2017) Department of Economic and Social Affairs, Population Division. World population prospects: the 2017 revision, methodology of the United Nations population estimates and projections, Working Paper No. ESA/P/WP.250. United Nations, New York
2. OMS (2015) https://www.un.org/sustainabledevelopment/es/objetivos-de-desarrollo-sostenible/
3. Ranganathan J, Waite R, Searchinger T, Hanson C (2018) How to sustainably feed 10 billion people by 2050, in 21 charts. World Resources Institute, Washington, DC

4. Shang L, Heckelei T, Gerullis MK, Börner J, Rasch S (2021) Adoption and diffusion of digital farming technologies-integrating farm-level evidence and system interaction. Agric Syst 190: 103074
5. Burra DD, Hildebrand J, Giles J, Nguyen T, Hasiner E, Schroeder K, Treguer D, Juergenliemk A, Horst A, Jarvis A, Kropf W (2021) Digital agriculture profile: Viet Nam. Rome, FAO, 22p
6. Kumar P, Chouriya A (2023) Chapter 1: Advances in precision farming: integrating AI and IoT technologies. In: Polara AM, Sahoo S, Jolly GE, Chennalwad SP, Asodariya KB (eds) Novel approaches in agronomy (vol 2). New Delhi (India), Elite Publishing House, pp 1–13
7. Elbasi E, Mostafa N, AlArnaout Z, Zreikat AI, Cina E, Varghese G, Shdefat A, Topcu AE, Abdelbaki W, Mathew S, Zaki C (2023) Artificial intelligence technology in the agricultural sector: a systematic literature review. IEEE Access 11:171–202
8. Bąba W, Kompała-Bąba A, Zabochnicka-Świątek M, Luźniak J, Hanczaruk R, Adamski A, Kalaji HM (2019) Discovering trends in photosynthesis using modern analytical tools: more than 100 reasons to use chlorophyll fluorescence. Photosynthetica 57:668–679. https://doi.org/10.32615/ps.2019.069
9. Goltsev V, Zaharieva I, Chernev P, Kouzmanova HM, Kalaji IY, Krasteva V, Alexandrov V, Stefanov D, Allakhverdiev SI, Strasser RJ (2012) Drought-induced modifications of photosynthetic electron transport in intact leaves: analysis and use of neural networks as a tool for a rapid non-invasive estimation. BBA-Bioenergetics 1817:1490–1498. https://doi.org/10.1016/j.bbabio.2012.04.018
10. Kalaji HM, Bąba W, Gediga K, Goltsev V, Samborska IA, Cetner MD, Dimitrova S, Piszcz U, Bielecki K, Karmowska K, Dankov K, Kompała-Bąba A (2017) Chlorophyll fluorescence as a tool for nutrient status identification in rapeseed plants. Photosynth Res 136:329–343. https://doi.org/10.1007/s11120-017-0467-7
11. Feng P, Wang B, Liu DL, Waters C, Xiao D, Shi L, Yu Q (2020) Dynamic wheat yield forecasts are improved by a hybrid approach using a biophysical model and machine learning technique. Agric For Meteorol 285–286:107922. https://doi.org/10.1016/j.agrformet.2020.107922
12. Guo Y, Fu Y, Hao F, Zhang X, Wu W, Jin X, Bryant CR, Senthilnath J (2021) Integrated phenology and climate in rice yields prediction using machine learning methods. Ecol Indic 120:106935. https://doi.org/10.1016/j.ecolind.2020.106935
13. Feng P, Wang B, Liu DL, Waters C, Yu Q (2019) Incorporating machine learning with biophysical model can improve the evaluation of climate extremes impacts on wheat yield in south-eastern Australia. Agric For Meteorol 275:100–113. https://doi.org/10.1016/j.agrformet.2019.05.018
14. Cao J, Zhang Z, Tao F, Zhang L, Luo Y, Zhang J, Han J, Xie J (2021) Integrating multi-source data for rice yield prediction across China using machine learning and deep learning approaches. Agric For Meteorol 297:108275. https://doi.org/10.1016/j.agrformet.2020.108275
15. Cai Y, Guan K, Peng J, Wang S, Seifert C, Wardlow B, Li Z (2018) A high-performance and in-season classification system of field-level crop types using time-series Landsat data and a machine learning approach. Remote Sens Environ 210:35–47. https://doi.org/10.1016/j.rse.2018.02.045
16. Khanal S, Fulton J, Klopfenstein A, Douridas N, Shearer S (2018) Integration of high resolution remotely sensed data and machine learning techniques for spatial prediction of soil properties and corn yield. Comput Electron Agric 153:213–225. https://doi.org/10.1016/j.compag.2018.07.016
17. Peng B, Guan K, Zhou W, Jiang C, Frankenberg C, Sun Y, He L, Kohler P (2020) Assessing the benefit of satellite-based solar-induced chlorophyll fluorescence in crop yield prediction. Int J Appl Earth Obs Geoinf 90:102126. https://doi.org/10.1016/j.jag.2020.102126
18. Kang Y, Ozdogan M, Zhu X, Ye Z, Hain C, Anderson M (2020) Comparative assessment of environmental variables and machine learning algorithms for maize yield prediction in the US Midwest. Environ Res Lett 15:064005. https://doi.org/10.1088/1748-9326/ab7df9

19. Tian H, Wang P, Tansey K, Zhang J, Zhang S, Li H (2021) An LSTM neural network for improving wheat yield estimates by integrating remote sensing data and meteorological data in the Guanzhong Plain, PR China. Agric For Meteorol 310:108629. https://doi.org/10.1016/j.agrformet.2021.108629
20. Liu Y, Wang S, Wang X, Chen B, Chen J, Wang J, Huang M, Wang Z, Ma L, Wang P, Amir M, Zhu K (2022) Exploring the superiority of solar-induced chlorophyll fluorescence data in predicting wheat yield using machine learning and deep learning methods. Comput Electron Agric 192:106612. https://doi.org/10.1016/j.compag.2021.106612
21. Fonseca de Oliveira GR, Mastrangelo CB, Hirai WY, Batista TB, Sudki JM, Petronilio ACP, Crusciol CAC, Amaral da Silva EA (2022) An approach using emerging optical technologies and artificial intelligence brings new markers to evaluate Peanut seed quality. Front Plant Sci 13:849986. https://doi.org/10.3389/fpls.2022.849986
22. Lu M, Gao P, Hu J, Hou J, Wang D (2023) A classification method of stress in plants using unsupervised learning algorithm and chlorophyll fluorescence technology. Front Plant Sci 14: 1202092. https://doi.org/10.3389/fpls.2023.1202092
23. Park B, Wi S, Chung H, Lee H (2024) Chlorophyll fluorescence imaging for environmental stress diagnosis in crops. Sensors 24:1442. https://doi.org/10.3390/s24051442
24. Khruschev SS, Plyusnina TY, Antal TK, Pogosyan SI, Riznichenko GY, Rubin AB (2022) Machine learning methods for assessing photosynthetic activity: environmental monitoring applications. Biophys Rev 14:821–842. https://doi.org/10.1007/s12551-022-00982-2
25. Zavafer A, Labeeuw L, Mancilla C (2020) Global trends of usage of chlorophyll fluorescence and projections for the next decade. Plant Phenomics 2020. https://doi.org/10.34133/2020/6293145
26. Suggett DJ, Oxborough K, Baker NR, MacIntyre HL, Kana TM, Geider RJ (2003) Fast repetition rate and pulse amplitude modulation chlorophyll a fluorescence measurements for assessment of photosynthetic electron transport in marine phytoplankton. Eur J Phycol 38:371–384. https://doi.org/10.1080/0967026031000161265
27. Nishina H (2015) Development of speaking plant approach technique for intelligent greenhouse. Agric Agric Sci Procedia 3:9–13. https://doi.org/10.1016/j.aaspro.2015.01.004
28. Takayama K, Hirota R, Takahashi N, Nishina H, Arima S, Yamamoto K, Sakai Y, Okada H (2014) Development of chlorophyll fluorescence imaging robot for practical use in commercial greenhouse. ActaHortic 1037:671–676. https://doi.org/10.17660/ActaHortic.2014.1037.86
29. Newcomb M, Shakoor N (2022) Nighttime chlorophyll fluorescence imaging of dark-adapted plants using a robotic field phenotyping platform. In: Lorence A, Medina Jimenez K (eds) High-throughput plant phenotyping. Methods in molecular biology, vol 2539. Humana, New York. https://doi.org/10.1007/978-1-0716-2537-8_17
30. Peruzzi A, Frasconi C, Martelloni L, Fontanelli M, Raffaelli M (2012). A new prototype to perform precision treatments of activated steaming to kill weed seed before sowing of organic vegetable. In Proceedings of the First International Conference on Robotics and Associated High-technologies and Equipment for Agriculture. Applications of automated systems and robotics for crop protection in sustainable precision agriculture, (RHEA-2012) Pisa, Italy: Univ. of Pisa, September 19–21, 2012, pp 49–54
31. Visser R, Timmermans AJM (1996) Weed-it: a new selective weed control system. In: Optics in agriculture, forestry, and biological processing II, vol 2907. SPIE, Bellingham, pp 120–129
32. Dhanya VG, Subeesh A, Kushwaha NL, Vishwakarma DK, Kumar TN, Ritika G, Singh AN (2022) Deep learning based computer vision approaches for smart agricultural applications. Artif Intell Agric 6:211–229. https://doi.org/10.1016/j.aiia.2022.09.007
33. Bilbao A, Bilbao E, Ciancio N, Giménez V, Muguerza J, Robles Terán L, Tinghitella G (2022) Aplicaciones Selectivas: Experiencias de uso y análisis de la tecnología en la Argentina. Contenidos CREA. https://www.contenidoscrea.org.ar/agricultura/manuales/publicaciones/aplicaciones-selectivas-nuevo-manual-n5326100
34. Soliense (n.d.) Fluorescence/photosynthetic properties: LIFT-FRR method. https://soliense.com/LIFT_Method.php

35. Plant Phenotyping (2016) Plant Phenotyping Survey 2016. https://www.plant-phenotyping.org/ippn-survey_2016
36. Soliense (n.d.) Terrestrial LIFT instrument: LIFT-REM fluorometer. https://soliense.com/LIFT_Terrestrial.php
37. European Space Agency, Mission Flex. https://earth.esa.int/eogateway/missions/flex. Last accessed October 2, 2024
38. Soliense (n.d.) Marine LIFT instrument: LIFT-FRR fluorometer. https://soliense.com/LIFT_Marine.php
39. Gorbunov MY, Falkowski PG (2020) Using chlorophyll fluorescence kinetics to determine photosynthesis in aquatic ecosystems. Limnol Oceanogr 66:1–13. https://doi.org/10.1002/lno.11581
40. Gorbunov MY, Falkowski PG (2022) Using chlorophyll fluorescence to determine the fate of photons absorbed by phytoplankton in the world's oceans. Annu Rev Mar Sci. https://doi.org/10.1146/annurev-marine-032621-122346
41. Alonso L, Gomez-Chova L, Vila-Frances J, Amoros-Lopez J, Guanter L, Calpe J, Moreno J (2008) Improved fraunhofer line discrimination method for vegetation fluorescence quantification. IEEE Geosci Remote Sens Lett 5:620–624. https://doi.org/10.1109/LGRS.2008.2001180
42. Meroni M, Busetto L, Colombo R, Guanter L, Moreno J, Verhoef W (2010) Performance of spectral fitting methods for vegetation fluorescence quantification. Remote Sens Environ 114(2):363–374. https://doi.org/10.1016/j.rse.2009.09.010
43. Zhang G, Roslan SNAB, Shafri HZM, Zhao Y, Wang C, Quan L (2024) Predicting wheat yield from 2001 to 2020 in Hebei Province at county and pixel levels based on synthesized time series images of Landsat and MODIS. Sci Rep 14:16212. https://doi.org/10.1038/s41598-024-67109-3
44. Naethe P, De Sanctis A, Burkart A, Campbell PKE, Colombo R, Di Mauro B, Damm A, El-Madany T, Fava F, Gamon JA, Huemmrich KF, Migliavacca M, Paul-Limoges E, Rascher U, Rossini M, Schüttemeyer D, Tagliabue G, Zhang Y, Julitta T (2024) Towards a standardized, ground-based network of hyperspectral measurements: combining time series from autonomous field spectrometers with Sentinel-2. Remote Sens Environ 303:114013. https://doi.org/10.1016/j.rse.2024.114013
45. Morozumi T, Kato T, Kobayashi H, Sakai Y, Nakashima N, Buareal K, Nishida Nasahara K, Kawaguchi Akitsu T, Murayama S, Noda HM, Muraoka H (2023) Contributions of the understory and midstory to total canopy solar-induced chlorophyll fluorescence in a ground-based study in conjunction with seasonal gross primary productivity in a cool-temperate deciduous broadleaf forest. Remote Sens Environ 284:113340. https://doi.org/10.1016/j.rse.2022.113340
46. Romero JM, Torres R, Ospina Calvo B, Diz VE, Iriel A, Cordon GB, Lagorio MG (2020) Photochemistry and photophysics of biological systems: chlorophyll fluorescence and photosynthesis. Asociación Química Argentina; Anales de la Asociación Química Argentina 107(2): 1–32
47. Omasa K, Konishi A, Tamura H, Hosoi F (2009) 3D confocal laser scanning microscopy for the analysis of chlorophyll fluorescence parameters of chloroplasts in intact leaf tissues. Plant Cell Physiol 50(1):90–105. https://doi.org/10.1093/pcp/pcn174
48. Jonkman J, Brown CM (2015) Any way you slice it-a comparison of confocal microscopy techniques. J Biomol Tech 26(2):54–65. https://doi.org/10.7171/jbt.15-2602-003
49. Tseng YC, Chu SW (2017) High spatio-temporal-resolution detection of chlorophyll fluorescence dynamics from a single chloroplast with confocal imaging fluorometer. Plant Methods. https://doi.org/10.1186/s13007-017-0194-2
50. Storti M, Hsine H, Uwizeye C, Bastien O, Yee DP, Chevalier F, Decelle J, Giustini C, Béal D, Curien G, Finazzi G, Tolleter D (2023) Tailoring confocal microscopy for real-time analysis of photosynthesis at single-cell resolution. Cell Rep Methods 3(9):100568. https://doi.org/10.1016/j.crmeth.2023.100568

51. Rammler T, Wackenhut F, Rapp J, zur Oven-Krockhaus S, Forchhammer K, Meixner AJ, Harter K (2023) Analysis of fast fluorescence kinetics of a single cyanobacterium trapped in an optical microcavity. Plan Theory 12(3):607. https://doi.org/10.3390/plants12030607
52. Küpper H, Benedikty Z, Morina F, Andresen E, Mishra A, Trtílek M (2019) Analysis of OJIP chlorophyll fluorescence kinetics and QA reoxidation kinetics by direct fast imaging. Plant Physiol 179(2):369–381. https://doi.org/10.1104/pp.18.00953
53. Sánchez-Moreiras AM, Graña E, Reigosa MJ, Araniti F (2020) Imaging of chlorophyll a fluorescence in natural compound-induced stress detection. Front Plant Sci 11:583590. https://doi.org/10.3389/fpls.2020.583590
54. Arief MAA, Kim H, Kurniawan H, Nugroho AP, Kim T, Cho BK (2023) Chlorophyll fluorescence imaging for early detection of drought and heat stress in strawberry plants. Plan Theory 12(6):1387. https://doi.org/10.3390/plants12061387
55. Legendre R, Basinger NT, van Iersel MW (2021) Low-cost chlorophyll fluorescence imaging for stress detection. Sensors 21(6):2055. https://doi.org/10.3390/s21062055
56. Colussi A, Bokhari SNH, Mijovilovich A, Koník P, Küpper H (2024) Acclimation to medium-level non-lethal iron limitation: adjustment of electron flow around the PSII and metalloprotein expression in *Trichodesmium erythraeum* IMS101. Biochim Biophys ActaBioenerg 1865(1): 149015. https://doi.org/10.1016/j.bbabio.2023.149015
57. Herppich WB (2021) Chlorophyll fluorescence imaging for process optimisation in horticulture and fresh food production. Photosynthetica 59:62–77. https://doi.org/10.32615/ps.2021.033
58. Rühle T, Reiter B, Leister D (2018) Chlorophyll fluorescence video imaging: a versatile tool for identifying factors related to photosynthesis. Front Plant Sci 9:55. https://doi.org/10.3389/fpls.2018.00055
59. Valcke R (2021) Can chlorophyll fluorescence imaging make the invisible visible? Photosynthetica 59(Special Issue):381–398. https://doi.org/10.32615/ps.2021.017
60. Moustaka J, Moustakas M (2023) Early-stage detection of biotic and abiotic stress on plants by chlorophyll fluorescence imaging analysis. Biosensors 13(8):796. https://doi.org/10.3390/bios13080796
61. Pérez-Bueno ML, Pineda M, Barón M (2019) Phenotyping plant responses to biotic stress by chlorophyll fluorescence imaging. Front Plant Sci 10:1135. https://doi.org/10.3389/fpls.2019.01135
62. Oláh V, Hepp A, Irfan M, Mészáros I (2021) Chlorophyll fluorescence imaging-based duckweed phenotyping to assess acute phytotoxic effects. Plan Theory 10(12):2763. https://doi.org/10.3390/plants10122763
63. Deng Y, Xin N, Zhao L, Shi H, Deng L, Han Z, Wu G (2024) Precision detection of salt stress in soybean seedlings based on deep learning and chlorophyll fluorescence imaging. Plan Theory 13(15):2089. https://doi.org/10.3390/plants13152089
64. Zhou C, Le J, Hua D, He T, Mao J (2019) Imaging analysis of chlorophyll fluorescence induction for monitoring plant water and nitrogen treatments. Measurement 136:478–486. https://doi.org/10.1016/j.measurement.2018.12.088
65. Zhang C, Zhou L, Xiao Q, Bai X, Wu B, Wu N, Zhao Y, Wang J, Feng L (2022) End-to-end fusion of hyperspectral and chlorophyll fluorescence imaging to identify Rice stresses. Plant Phenomics 2022:9851096. https://doi.org/10.34133/2022/9851096
66. Vítek P, Novotná K, Hodaňová P, Rapantová B, Klem K (2017) Detection of herbicide effects on pigment composition and PSII photochemistry in Helianthus annuus by Raman spectroscopy and chlorophyll a fluorescence. Spectrochim Acta A Mol Biomol Spectrosc 170:234–241. https://doi.org/10.1016/j.saa.2016.07.025
67. Vítek P, Veselá B, Klem K (2020) Spatial and temporal variability of plant leaf responses cascade after PSII inhibition: Raman, chlorophyll fluorescence and infrared thermal imaging. Sensors 20(4):1015. https://doi.org/10.3390/s20041015
68. Long Y, Ma M (2020) Recognition of drought stress state of tomato seedling based on chlorophyll fluorescence imaging. IEEE Access 10:48633–48642. https://doi.org/10.1109/ACCESS.2022.3168862

69. Dong Z, Men Y, Liu Z, Li J, Ji J (2019) Application of chlorophyll fluorescence imaging technique in analysis and detection of chilling injury of tomato seedlings. Comput Electron Agric 168:105109. https://doi.org/10.1016/j.compag.2019.105109
70. Li H, Wang P, Weber JF, Gerhards R (2017) Early identification of herbicide stress in soybean (Glycine max (L.) Merr.) using chlorophyll fluorescence imaging technology. Sensors 18(1):21. https://doi.org/10.3390/s18010021
71. Wang P, Peteinatos G, Li H, Gerhards R (2016) Rapid in-season detection of herbicide resistant *Alopecurus myosuroides* using a mobile fluorescence imaging sensor. Crop Prot 89:170–177. https://doi.org/10.1016/j.cropro.2016.07.022
72. Wang P, Peteinatos G, Li H, Brändle F, Pfündel E, Drobny H, Gerhards R (2017) Rapid monitoring of herbicide-resistant *Alopecurus myosuroides* Huds. Using chlorophyll fluorescence imaging technology. J Plant Dis Prot 125:187–195. https://doi.org/10.1007/s41348-017-0131-7
73. Herritt MT, Pauli D, Mockler TC, Thompson AL (2020) Chlorophyll fluorescence imaging captures photochemical efficiency of grain sorghum (Sorghum bicolor) in a field setting. Plant Methods 16:109. https://doi.org/10.1186/s13007-020-00650-0
74. Herritt MT, Long JC, Roybal MD, Moller DC, Mockler TC, Pauli D, Thompson AL (2021) FLIP: fluorescence imaging pipeline for field-based chlorophyll fluorescence images. SoftwareX 14:100685. https://doi.org/10.1016/j.softx.2021.100685

Correction to: Chapters

María Gabriela Lagorio

Correction to:
Comprehensive Series in Photochemical and Photobiological Sciences, Fluorescence of Living Plants,
https://doi.org/10.1007/17361_2024_1
https://doi.org/10.1007/17361_2024_2
https://doi.org/10.1007/17361_2024_3
https://doi.org/10.1007/17361_2025_4

The chapters were inadvertently published with an incorrect copyright holder name. The copyright holder name for these chapters has been updated as follows:

European Society for Photobiology, under exclusive license to Springer Nature Switzerland AG

The updated versions of these chapters can be found at
https://doi.org/10.1007/17361_2024_1
https://doi.org/10.1007/17361_2024_2
https://doi.org/10.1007/17361_2024_3
https://doi.org/10.1007/17361_2025_4

M. G. Lagorio (ed.), Comprehensive Series in Photochemical and Photobiological Sciences, Fluorescence of Living Plants 1, https://doi.org/10.1007/17361_2025_13
Published online: 19 August 2025

MIX
Papier aus verantwortungsvollen Quellen
Paper from responsible sources
FSC® C105338

If you have any concerns about our products, you can contact us on
ProductSafety@springernature.com

In case Publisher is established outside the EU, the EU authorized representative is:
Springer Nature Customer Service Center GmbH
Europaplatz 3, 69115 Heidelberg, Germany

Printed by Libri Plureos GmbH
in Hamburg, Germany